T0339863

Sustainable Hydropower
in West Africa

Sustainable Hydropower in West Africa

Planning, Operation, and Challenges

Edited by

AMOS KABO-BAH

CHUKWUEMEKA J. DIJI

Academic Press is an imprint of Elsevier
125 London Wall, London EC2Y 5AS, United Kingdom
525 B Street, Suite 1800, San Diego, CA 92101-4495, United States
50 Hampshire Street, 5th Floor, Cambridge, MA 02139, United States
The Boulevard, Langford Lane, Kidlington, Oxford OX5 1GB, United Kingdom

Library of Congress Cataloging-in-Publication Data
A catalog record for this book is available from the Library of Congress

British Library Cataloguing-in-Publication Data
A catalogue record for this book is available from the British Library

ISBN 978-0-12-813016-2

For information on all Academic Press publications
visit our website at https://www.elsevier.com/books-and-journals

Working together
to grow libraries in
developing countries

www.elsevier.com • www.bookaid.org

Publisher: Joe Hayton
Acquisition Editor: Raquel Zanol
Editorial Project Manager: Katie Chan
Production Project Manager: Surya Narayanan Jayachandran
Cover Designer: Christian J. Bilbow

Typeset by SPi Global, India

Contents

Contributors   xi
Foreword   xiii
Introduction   xv

1. Climate Change and Hydrovision

*Derrick M.A. Sowa, Basil Amuzu-Sefordzi, Thelma D. Baddoo,
Mark Amo-Boateng, Martin K. Domfeh*

1	Introduction	1
2	The Current State and Prospects of Hydropower Development in Africa	4
3	Climate Change and Energy Production in Ghana	5
	3.1 Country Profile and Electric Power Production	5
	3.2 Climate Change Impacts	11
	3.3 Alternative Energy Sources—Renewable Energy	11
4	Marine Energy Conversion Systems	15
	4.1 Prototype (Initial Design) of a Novel Marine Energy Conversion System	16
	4.2 Compatibility of the System in Ghana	18
5	Conclusion	23
	References	24

2. Promoting Research in Sustainable Energy in Africa—The CIRCLE Model

Benjamin Apraku Gyampoh

1	Introduction	29
2	The CIRCLE Approach	30
	2.1 Supporting Different Needs of ECRs	30
	2.2 Intra-African Collaboration	31
	2.3 Keeping the Links With the North	31
	2.4 Creating a Conducive Environment for Continuous Researcher Development	31
	2.5 Research Uptake	32
3	Driving Solutions Through Intra-African Research Collaboration	32
4	CIRCLE-Funded Research in Energy	33

5 Climate Change, Hydropower, and Energy in Africa 34
6 Research Uptake and the Hydropower Energy Conference 34
7 Conclusion 35
References 35

3. Hydropower and the Era of Climate Change and Carbon Financing: The Case From Sub-Saharan Africa

Amos T. Kabo-bah, Caleb Mensah

1 Introduction 37
2 Hydropower and Climate Change 39
   2.1 Carbon Financing 42
   2.2 Hydropower as a Clean Development Mechanism Tool 43
   2.3 Carbon Markets and Hydropower 44
3 The Case of Ghana 45
4 Conclusions 47
5 Recommendations 48
References 49
Further Reading 51

4. Hydropower Development—Review of the Successes and Failures in the World

Xie Yuebo, Amos T. Kabo-bah, Kamila J. Kabo-bah, Martin K. Domfeh

1 Introduction 53
2 Methodology of the Study 54
3 Successful and Failed Hydropower Projects 54
4 Lessons From Cited Case Studies 58
5 Conclusion 60
References 60

5. Climate Change and Societal Change—Impact on Hydropower Energy Generation

Mary Antwi, Daniella D. Sedegah

1 Introduction 63
2 Climate Change and Hydropower Generation of Electricity in Sub-Saharan Africa 65
3 Climate Change and Patterns of Societal Consumption of Generated Hydropower 66
   3.1 Availability of Electrical Energy Supply in Sub-Saharan Africa 66
4 Climate Change and Societal Change in the Face of Hydropower Energy Generation 69
5 Conclusion 71
References 71

6. Renewable Energy and Sustainable Development

Samuel Gyamfi, Nana S.A. Derkyi, Emmanuel Y. Asuamah, Israel J.A. Aduako

1	Introduction	75
2	Geographic and Climatic Conditions of West Africa	76
3	Renewable Energy, a Necessity for Sustainable Development	78
4	Renewable Energy and Sustainable Development	80
	4.1 Connection Between Renewable Energy and Sustainable Development	81
	4.2 Sustainability Indicators for Renewable Energy	82
5	Renewable Energy Potential in West Africa	82
	5.1 Theoretical Potential	82
	5.2 Geographic Potential	82
	5.3 Technical Potential	83
	5.4 Economic Potential	83
	5.5 Solar Photovoltaic (PV) and Concentrated Solar Power	83
6	Wind Energy Potentials in West Africa	85
7	Biomass for Energy Potential in West Africa	85
8	Barriers to Renewable Energy Integration	88
9	Observations and Discussions	89
10	Policy Recommendations and Way Forward	90
11	Conclusion	91
	References	92
	Further Reading	93

7. The Potential and the Economics of Hydropower Investment in West Africa

Samuel Gyamfi, Nana S.A. Derkyi, Emmanuel Y. Asuamah

1	Introduction	95
	1.1 Layout of West Africa	97
2	Hydropower Potential in West Africa	97
	2.1 Niger River	98
	2.2 Senegal River	98
	2.3 Volta River	99
3	Climate Change Uncertainties and the Prospects of Hydropower in West Africa	99
4	Economics of Hydropower in West Africa	101
	4.1 Cost of Hydropower Project	102
5	Economic Benefits of Hydroelectric Power Generation Over Other RE Sources	103
	5.1 The Base Load Power Case	104
6	Factors Affecting Private Sector Investment	104
7	Observations and Discussions	105
8	Conclusions	105
	References	106
	Further Reading	107

8. **Hydropower Generation and Its Related Impacts on Aquatic Life (Fisheries)**

   Berchie Asiedu, Francis E.K. Nunoo, Elliot H. Alhassan, Patrick K. Ofori-Danson

1	Introduction	109
2	Importance of Fisheries to the Ghanaian Economy	110
	2.1 Content of the Paper	110
3	Materials and Methods	110
	3.1 Study Area	110
	3.2 Document Analysis	113
4	Analysis of Impacts of Hydropower Development on Fish and Fisheries	113
	4.1 Impacts of Impoundments on Riverine Ecosystems	113
	4.2 Anticipated Impacts of the Bui Dam on Biodiversity	114
	4.3 Food Security and Fish Consumption	115
	4.4 Fish Diversity	116
	4.5 Fish Migration	116
	4.6 Fishers' Migration	116
	4.7 Fishing Effort	116
	4.8 Other Social and Economic Impacts	117
5	Conclusion	117
	References	117

9. **Socioeconomic Impacts of the Bui Hydropower Dam on the Livelihood of Women and Children**

   Nina Schafer, Heidi Megerle, Amos T. Kabo-bah

1	Introduction	121
	1.1 Bui Hydropower Dam	122
2	Methodology	124
3	Results	125
	3.1 Changes in Livelihood	127
	3.2 Water Situation	129
	3.3 Waste Management	131
	3.4 Lost Tourism in the Bui National Park	133
4	Discussion	133
5	Conclusion	135
	References	136

10. **Peri-urban Households' Constraints to Water Security and Changing Economic Needs Under Climate Variability in Ghana**

   Divine O. Appiah, Patrick Benebere, Felix Asante

1	Introduction	137
2	Methodology	139
	2.1 Profile of the Study Area	139
	2.2 Research Design	140

3 Results and Discussion   143
   3.1 Changing Economic Needs   143
   3.2 Local People's Perception of Climate Variability   146
   3.3 Adaptation to Climate Variability   151
   3.4 Constraints for Adaptation to Water Insecurity Under
        Climate Variability   154
4 Conclusion and Recommendation   155
   Acknowledgments   156
   References   156

11. Legislation on Hydropower Use and Development

Lilian Idiaghe

1 Introduction   159
2 The International Landscape   162
3 Hydropower in the ECOWAS Region   165
4 National Frameworks for Hydropower Use and Development   168
   4.1 Ghana   168
   4.2 Nigeria   171
   4.3 Sierra Leone   175
   4.4 Liberia   176
   4.5 Gambia   178
   4.6 Cape Verde   180
   4.7 Mali   181
   4.8 Senegal   183
5 Conclusion and Recommendations   184
   References   186

12. Re-engineering Hydropower Plant for Improved Performance

Eric A. Ofosu, Mark Amo-Boateng, Martin K. Domfeh, Robert Andoh

1 Introduction   189
2 The Case Study of the Akosombo Dam   190
3 Reoperation and Reoptimization of the Akosombo Dam   191
4 Optimizing Intake Configurations to Facilitate Power
   Generation at Low Water Levels at Akosombo Dam   193
5 Turbine Plant Intake Optimization   193
   References   194

13. Hydropower Generation in West Africa—The Working Solution Manual

Amos T. Kabo-bah, Chukwuemeka J. Diji, Kofi A. Yeboah

1 Introduction   197
   1.1 Importance of Hydropower in the World, Africa, and Ghana   197
2 Challenges   200
   2.1 Climate   200
   2.2 Infrastructure   201

2.3 Operation and Maintenance (O&M) 201
2.4 Life Cycles 201
2.5 Extreme Event-Dam failure 201
3 **Proposed Working Manual** 202
3.1 Climate 202
3.2 Infrastructure 202
3.3 Operation and Maintenance—Use of Robotics
and Digital Innovations, etc 203
3.4 Life-Cycle Integration With Other Sources of Energy 203
3.5 Water-Energy Nexus 204
3.6 Extreme Event—Preventing failure 204
3.7 Other Considerations 205
4 **Conclusion** 205
References 206

Index 209

Contributors

Numbers in parenthesis indicate the pages on which the authors' contributions begin.

Israel J.A. Aduako (75), University of Energy and Natural Resources (UENR), Sunyani, Ghana

Elliot H. Alhassan (109), University for Development Studies, Tamale, Ghana.

Mark Amo-Boateng (1, 189), University of Energy and Natural Resources, Sunyani, Ghana

Basil Amuzu-Sefordzi (1), Hohai University, Nanjing, China

Robert Andoh (189), AWD Consult. Inc., Portland, ME, United States

Mary Antwi (63), University of Energy and Natural Resources, Sunyani, Ghana

Divine O. Appiah (137), Kwame Nkrumah University of Science and Technology, Kumasi, Ghana

Felix Asante (137), Kwame Nkrumah University of Science and Technology, Kumasi, Ghana

Berchie Asiedu (109), University of Energy and Natural Resources, Sunyani, Ghana.

Emmanuel Y. Asuamah (75, 95), University of Energy and Natural Resources (UENR), Sunyani, Ghana

Thelma D. Baddoo (1), Hohai University, Nanjing, China

Patrick Benebere (137), Kwame Nkrumah University of Science and Technology, Kumasi, Ghana

Patrick K. Ofori-Danson (109), University of Ghana, Accra, Ghana

Nana S.A. Derkyi (75, 95), University of Energy and Natural Resources (UENR), Sunyani, Ghana

Chukwuemeka J. Diji (197), University of Ibadan, Nigeria

Martin K. Domfeh (1, 53, 189), University of Energy and Natural Resources, Sunyani, Ghana

Samuel Gyamfi (75, 95), University of Energy and Natural Resources (UENR), Sunyani, Ghana

Benjamin Apraku Gyampoh (29), African Academy of Sciences (AAS), Nairobi, Kenya

Lilian Idiaghe (159), University of Ibadan, Ibadan, Nigeria

Amos T. Kabo-bah (53, 121, 197), University of Energy and Natural Resources, Sunyani, Ghana; Hochschule für Forstwirtschaft, Rottenburg, Germany

Kamila J. Kabo-bah (53), University of Energy and Natural Resources, Sunyani, Ghana

Heidi Megerle (121), Hochschule für Forstwirtschaft, Rottenburg, Germany

Caleb Mensah (37), Earth Observation Research and Innovation Centre, University of Energy and Natural Resources, Sunyani, Ghana

Francis E.K. Nunoo (109), University of Ghana, Accra, Ghana

Eric A. Ofosu (189), University of Energy and Natural Resources, Sunyani, Ghana

Nina Schafer (121), Hochschule für Forstwirtschaft, Rottenburg, Germany

Daniella D. Sedegah (63), University of Energy and Natural Resources, Sunyani, Ghana

Derrick M.A. Sowa (1), Hohai University, Nanjing, China

Kofi A. Yeboah (197), University of Energy and Natural Resources, Sunyani, Ghana

Xie Yuebo (53), Hohai University, Nanjing, China

Foreword

Hydropower is the largest renewable source of electricity generation. Estimates show that about two-thirds of the world's economically feasible potential of hydropower is yet to be exploited. In Sub-Saharan Africa where 25% of people have access to electricity, the development of hydropower could become a valuable contribution to ensure that African countries meet the universal access to electricity by 2050. Notwithstanding, the impact of climate change on the hydrological cycle including its negative effects on socioeconomic livelihood of communities has also raised concerns in recent times.

In the West Africa subregion, it is expected that electricity demand will continue to increase in the coming decades, but one way to support the increasing demand is by investing more into renewable energies such as Hydropower Energy.

The book has carefully examined some of the successes and failures of hydropower in the World, analyzing some of the socioeconomic impacts of such schemes, economic potentials for new hydropower plants in the region, and in that end, summarizes some of the key blueprint innovative solutions that can be implemented for existing hydropower schemes. In addition, it presents a working manual for the establishment of new hydropower schemes in West Africa.

The book is a one stop for anyone interested in boosting the energy supply in West Africa and similar regions in the world and highly recommended for the policymaker, the engineer, the economist, and the community developer.

Esi Awuah
University of Energy and Natural Resources, Sunyani, Ghana

Introduction

Daniel Obeng-Ofori

University of Energy and Natural Resources, Sunyani, Ghana
Catholic University of Ghana, Sunyani, Ghana

Global energy security is a key challenge recently, due to high energy use fueled by rapid economic and population growth, the depletion of fossil fuel resources, and political instability in fossil fuel-rich areas (Kowalski and Vilogorac, 2008). This challenge has been reinforced by the conflicts in the Middle East in early 2011. Electricity generation is important for powering economic growth. However, climate change alters disaster risks for electricity generation and these considerations have been overlooked in energy research, policy, and planning. Climate change can exacerbate existing disaster risks and can increase the frequency and severity of some extreme climate events, such as heat waves, heavy precipitation events, and storms (IPCC, 2007). It is important to determine the links between changing disasters risks due to climate change and their impacts on electricity generation. The vulnerability of various electricity generation options such as fossil fuels, nuclear power, hydropower, and renewable needs to be assessed. Furthermore, it is useful to understand how changing disaster risks could be taken into account in electricity generation.

Climate Change has been used exhaustively in science, politics, and economics as a result of its related causes and impacts on our lives and livelihoods. Climate change may be explained as a significant statistical variation in the state of the climate or its variation over an extended period of decades or longer such that it significantly affects our lifestyle and livelihoods (Füssel and Klein, 2006).

Climate Change needs an enabling environment to operate/impact on. This enabling environment is termed a system. Various systems across the globe are expected to experience different degrees of effects. The nature and the extent to which a system gets exposed or affected by a mean state of change in climate is called exposure. The degree to which a natural or a human system may be adversely or beneficially influenced by climate-related stimuli either directly or indirectly is termed its sensitivity. Impact describes the resultant effects of climate change on human and natural systems. Impacts can also be categorized as potential and residual impacts. According to the United Nations (2004), there are four groups of vulnerability factors. These are physical, economic, social, and environmental factors. Physical factors describe the exposure of vulnerable elements within a region; economic factors describe the economic resources of individuals, population, groups, and communities; social factors describe noneconomic factors that determine the well-being of individuals, populations, groups, and communities, such as the level of education, security, access to basic human rights, and good governance; and environmental factors, which describe the state of the environment within a region.

The impacts of climate change across all countries in the world are well known and some of the threats such as sea level rise, floods, and droughts are already being experienced. For instance, studies conducted by the Council for Scientific and Industrial Research—Water Research Institute (CSIR-WRI) showed that Ghana even without the impact of the climate change is expected to become water stressed by 2025. In the case of the adverse effects of climate change, this implies that climate change would worsen the conditions of water scarcity in most parts across Ghana. The report from CSIR-WRI further stressed that there would be a general reduction in river flows by 15%–20% from 2020 and 30%–40% by 2050. There would a reduction in hydropower generation to about 60% for the year 2020, and from the year 2020, the country would experience acute water shortage (Kankam-Yeboah et al., 2011).

Ghana is already experiencing power cuts as a result of the fluctuating energy supplies in the country. The country has also heavily invested in hydropower dams including the Akosombo dam, the Kpong dam, and the recently constructed Bui dam. Generally, the performance of these hydropower plants is directly based on the water levels in their reservoirs. However, the changing climate conditions in the catchments for these dams severely affect their performance. With the rapid population growth, rising commercial and industrial activity, and urbanization in the country, the energy demand of the country is bound to increase.

Hydropower is the most commonly used renewable energy source today. The World Energy Council estimates that there is a global potential of more than 41,202 TWh/year (4703 GW) of hydropower with a technically exploitable potential of more than 16,494 TWh/year (1883 GW) (WEC, 2007). Both large hydropower and small hydropower play a role in the energy provision of many developing countries and some developed countries (e.g., Norway, Sweden). Large hydropower plants usually have a capacity of more than 10 MW and are linked to dams and reservoirs, whereas small hydropower plants have lower capacities and are built around river run-offs with low environmental impact.

In West Africa, there has been rising demand for power due to rapid economic growth, population growth, and urbanization. Hydropower is a major and vital source of energy for meeting this increasing electricity demand particularly in the industrial, domestic, and service sectors of the economy. Sustainable electricity supply not only supports social and economic development processes but also environmental and global climate change management and hence its importance in attainment of the Sustainable Development Goals. The West Africa region has great hydropower potential that can provide for the energy requirements of the region in an affordable, sustainable, and secure manner, regardless of the existing challenges of climate variability. A long-term vision is therefore needed to make the best use of the available domestic resources, given the long-lasting nature of hydropower energy infrastructure. However, if this electricity is generated unsustainably and the trend is not controlled, it could aggravate the environmental and climate change management problems that the world is currently experiencing.

Climate change and its impacts on changing disaster risks pose another major challenge to the energy sector and particularly electricity generation, which is important for powering development and economic growth. There are two major ways in which climate change and intensified disaster risks can affect the hydropower sector: First, hydropower plants are highly vulnerable to water stress, drought, decreased precipitation, and higher temperatures. This can lead to water discharge, which results in reduced electricity production. Second, hydropower plants could be impacted by changing disaster risks. This may include the increased intensity and frequency of storms, floods, and heavy precipitation events.

Hydropower is highly vulnerable to climate change and changing disaster risks. Changes in river flow, evaporation, and dam safety are the main drivers through which climate change impacts hydrological processes (Mideksa and Kallbekken, 2010). The potential to generate electricity correlates strongly with the changes in water discharge. Although increased precipitation and river flow eventually boosts energy production, excess flow may impact negatively.

This book therefore examines the dependence between climate and hydropower and the need to develop innovative solutions in harnessing hydropower for development. The book also argues that research support and leadership skills are needed in order to maintain and develop new hydropower schemes. The book provides the working model of the Climate Impact Research Capacity and Leadership Enhancement in Sub-Saharan Africa (CIRCLE), which is an initiative of the UK's Department for International Development (DFID) to develop the skills and research output of early career African researchers in the field of climate change and its local impacts on development. Such initiatives are vital for generating new knowledge and leadership in the management and development of power industry in Africa.

The book provides some suggestions on how changing disaster risks could be taken into account in hydropower generation. These include among other things Assessing disaster risks in Environmental Impact Assessments, improving feasibility studies and siting procedures for new power plants, enhancing interministerial linkages in relation to energy policy and planning, and developing climate change adaptation strategies for hydropower generation. This approach can enhance the ability of developing country governments, private firms, and civil society organizations to build the resilience of communities to disasters and climate change with regard to hydropower generation.

The management and development of hydropower schemes require good knowledge of some of the successes and failures. The book therefore takes a bird view of some of these historical events and outlines some of the key lessons learnt, and what can be important for the addressing the challenges in the hydropower sector in West Africa. Carbon financing for hydropower is equally a good option for West Africa and the book points out some critical issues of consideration for policy makers within this region. It also critically examines the investment potentials and economic returns for hydropower development in West Africa. Such new investments in West Africa must also take into careful consideration the socioeconomic impacts, environmental impacts, and agricultural productivity. Recommendations on areas of consideration when it comes to new hydropower developments have been outlined for future considerations.

The book presents a working manual for practitioners, policymakers, and scientists with the interest of having a first-hand information of the hydropower sector within West Africa with key strategies for operation and management of existing schemes, and ideas for future sustainable investments.

REFERENCES

Füssel, H.M., Klein, R.J.T., 2006. Climate change vulnerability assessments: an evolution of conceptual thinking. Clim. Chang. 75 (3), 301–329. https://doi.org/10.1007/s10584-006-0329-3.

IPCC Intergovernmental Panel on Climate Change, 2007. IPCC Forth Assessment Report: Climate Change 2007: Synthesis Report. IPCC, Geneva.

Kankam-Yeboah, K., Amisigo, B., Obuobi, E., 2011. Climate change impacts on water resources in Ghana. http://natcomreport.com/ghana/livre/climate-change.pdf.

Kowalski, G., Vilogorac, S., 2008. Energy Security Risks and Risk Mitigation: An Overview. www.preventionweb.net/files/8066_Pagesfromannualreport2008.pdf.

Mideksa, T.K., Kallbekken, S., 2010. The impact of climate change on the electricity market: a review. Energy Policy 38 (7), 3579–3585.

United Nations, 2004. Living with Risk: A Global Review of Disaster Reduction Initiatives. United Nations International Strategy for Disaster Reduction, Geneva, Switzerland.

WEC World Energy Council, 2007. 2007 Survey of Energy Resources. www.worldenergy.org/documents/ser2007_final_online_version_1.pdf.

Chapter 1

Climate Change and Hydrovision

Derrick M.A. Sowa*, Basil Amuzu-Sefordzi*, Thelma D. Baddoo*,
Mark Amo-Boateng[†], Martin K. Domfeh[†]
*Hohai University, Nanjing, China
[†]University of Energy and Natural Resources, Sunyani, Ghana

Chapter Outline

1. **Introduction**	1	4. **Marine Energy Conversion**	
2. **The Current State and**		**Systems**	15
Prospects of Hydropower		4.1 Prototype (Initial Design)	
Development in Africa	4	of a Novel Marine Energy	
3. **Climate Change and Energy**		Conversion System	16
Production in Ghana	5	4.2 Compatibility of the	
3.1 Country Profile and		System in Ghana	18
Electric Power Production	5	5. **Conclusion**	23
3.2 Climate Change Impacts	11	**References**	24
3.3 Alternative Energy			
Sources—Renewable Energy	11		

1. INTRODUCTION

Demand for energy and related services, to meet socioeconomic development and improve human health and welfare, is growing. All societies have need of both energy resources and services to meet basic human needs and also facilitate productive practices (IPCC, 2011). The global energy consumption is estimated to rise significantly in the coming years (Crabtree et al., 2004; Edwards et al., 2008; McDowall and Eames, 2006). Most parts of Euro-Asia and The Americas (particularly, North America) have impressive electrification. In contrast, as many as two billion people worldwide lack electricity today (Flavin and O'Meara, 1997). IRENA (2013) has revealed that over 67% of sub-Saharan Africans lack access to electricity, and as rapid population growth in developing countries continues, the demand for electricity will almost certainly rise.

The adverse impacts of traditional and accustomed energy sources—oil, coal, and natural gas—are detrimental to economic progress, the environment, and sustainability of humans (Akella et al., 2009); and as dynamic as the world

Sustainable Hydropower in West Africa. https://doi.org/10.1016/B978-0-12-813016-2.00001-0

and technology is, there has been the subtle evolution of energy production from renewable resources. The potential of renewable energy sources is massive as they can standardly meet many times the world's energy demand (Akella et al., 2009). Different forms of renewable energy can supply electricity, thermal energy, and mechanical energy, as well as produce fuels that are able to fulfill multiple energy service needs (IPCC, 2011). Renewable energy has been deliberated as one of the strong contenders to improve the predicament of two billion people, mostly in rural regions, without access to modern forms of energy (Mundial, 1996; Painuly, 2001). Renewable energy sources such as solar, wind, geothermal, modern bioenergy, hydropower, and marine energy are currently contributing about two percent of the world's total energy demand and continue to increase as the costs of, for example, solar and wind power systems, have declined substantially in the past 30 years, and continue to drop, while the price of oil and gas keeps fluctuating (Akella et al., 2009).

Marine energy conversion technology is one of the renewable resources that have received widespread attention in different parts of the world. Although this technology is in its early stages, particularly ocean wave energy, the idea of energy extraction from the ocean is not new. Thorpe (1999) estimated that the potential worldwide wave power resource is 2 TW. At the present time, the economic prowess and competitiveness of this resource is nowhere near the traditional and more matured technologies such as wind energy, yet governments and policy makers are showing great interest in this industry due to its sustainability and low environmental impacts. Several countries like the United Kingdom, Denmark, and Norway have had substantial wave power resources and have been in the business of wave energy utilization with support from their respective governments. Accounts from Callaghan and Boud (2006) have demonstrated how much of the current UK electricity demand can be met via wave energy and tidal stream generation. This is in the region of about 20%, which is very significant.

Lack of sufficient and reliable electricity has been a problem that has confronted Ghana for several decades. With the rise in population and surge in urbanization, the demand for the already-insufficient electricity is growing and will continue to grow. The inability of the energy sector to meet the growing demand is usually as a result of the low tariff system (USAID, 1999) and the failure to expand the energy capacity to meet the growing demand (ISSER, 2005). This inadequate power supply has incited erratic power outages—dubbed *Dumsor,* which means "switch off-switch on"—which pose serious problems to several households as well as small- and medium-scale enterprises (SMEs) that contribute significantly to the nation's economic development (Ayanda, 2011; Mensah, 2004). Reports from the Institute of Statistical, Social and Economic Research (ISSER, 2005) reveal that extra power generated from thermal plants purposely for augmenting electricity supply in Ghana has not provided a remedy to the inadequate and unreliable supply of electricity. Further, the electricity tariffs (Keener and Banerjee, 2005) partially impede the Electricity Company of

Ghana (ECG) in their attempts to change old and possibly worn-out transformers and loaders with modern ones of higher capacities. That notwithstanding, great efforts are being made to generate adequate electricity for the entire nation. In so doing, a new hydroelectric plant has been constructed on the Black Volta River at the Bui Gorge (i.e., the Bui dam). This is a 400-MW hydroelectric project, but currently producing about 90 MW of electricity (Boateng, 2014). In addition, a 111-km gas pipeline is also under construction from Atuabo where the Gas Processing Plant (GPP) is to be located to Aboadze (Boadu, 2014). This is a measure to provide the fuel source for the Volta River Authority's (VRA) gas power plant to produce electricity.

It has been observed that majority of Ghana's energy demand is supplied majorly by hydropower with fuel sources being secondary. The main challenge with these sources of energy generation is that the hydropower sources rely mainly on the *volume* and *discharge* of water from hydrological resources (e.g., rivers), which in turn depend on rainfall for their sustenance. Erratic precipitation coupled with unpredictable amounts of rainfall have caused severe fluctuation in the volume and discharge of water from most dams (particularly, the Akosombo Dam in Ghana). With the anticipated impacts of climate change on hydrological resources, this problem may further be worsened. The impacts of climate change on water resources have been a worldwide concern for some years now (Arnell, 1999; Vörösmarty et al., 2000; Christensen et al., 2004; Scanlon et al., 2007; Solomon, 2007; Bates et al., 2008; Piao et al., 2010; Chevallier et al., 2011; Hagemann et al., 2013), and the anticipated impacts on hydropower production have been broadly delved into (see Markoff and Cullen, 2008; Mideksa and Kallbekken, 2010; Hamududu and Killingtveit, 2012; Gaudard et al., 2013). With regard to the thermal sources of energy production, the fuel source—oil and or gas—is expensive, and barely sustainable considering Ghana's economy.

Other sources of renewable energy that are not particularly developed in Ghana need critical attention in order to effectively meet the ever-growing energy demand in the country. Ghana is fortunate to be endowed with several natural resources that can be tapped to generate her own electricity: the sun, wind, and the ocean. Similar to the different sources of renewables, marine energy has received considerable attention around the world, and ocean waves have been demonstrated to contain higher energy densities (Mccullen et al., 2002).

The magnitude of the energy deficit in Ghana coupled with the impacts of climate change on energy resources calls for inquiries into alternative sources of sustainable renewable energy. This chapter therefore presents a prototype (i.e., the design phase) of a novel marine energy conversion system that makes use of both ocean currents and wave energy applicable in Ghana. With the anticipation of increased sea levels and relative changes in met-ocean conditions, the entire system is fixed to an offshore breakwater to perform the dual function of energy production and control of sediment transport along the coast.

2. THE CURRENT STATE AND PROSPECTS OF HYDROPOWER DEVELOPMENT IN AFRICA

According to IPCC (2007), most countries in the African continent are most vulnerable to climate change impacts due to weak climate change adaptation mechanisms and policies. Renewable energy sources provide a high potential in mitigating these climate change impacts as well as addressing energy shortfalls. Considering the untold benefits associated with hydropower, it is therefore not surprising that many countries such as Cameroon, Sudan, Zambia, and DR Congo have a greater portion of their energy generation mix emanating from hydropower (World Bank, 2013). The estimated hydropower installed capacity for the continent in 2009 was 102 TWh/year with a 1174 TWh/year technical potential output (IEA, 2013). Cole et al. (2014) estimated 525 MW and 1.46 TWh for the mean hydropower capacity and the mean production for the countries in the African continent. The study made use of IPPC climate change scenarios, continental river flow models, and other dam and hydrological data sources. The study also provided a spatial distribution of the hydropower and plant capacities (Fig. 1). The darker areas have higher hydropower capacity compared to the lighter areas. The dams are illustrated by circles and the size of the circle indicates the plant capacity. It could be inferred from Fig. 1 that Nigeria has the greatest number of large hydropower plant facilities. Also, much of the hydropower sites are concentrated in the Western, Eastern, and Southern parts of Africa. In the Northern part of the continent, Morocco leads in the number of hydropower sites with the other North African countries showing very little dependence on hydropower sources. The research study also revealed

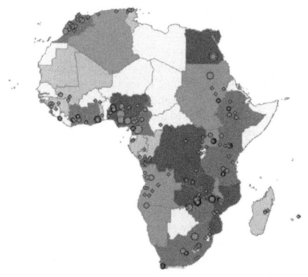

FIG. 1 Hydropower as a percentage of total capacity and future hydropower plants. Key: The darker the shade of a country, the greater its hydropower capacity. Dams are represented by circles with the size of the circle reflecting the capacity of the dam (Cole et al., 2014).

FIG. 2   River flow changes and current and future planned hydropower dams: increase *(blue)*, decrease *(red)*, no change *(orange)*. Key: Depicts for basins the ratio of the standard deviation of river flow of future relative to the past: 50% above = *red*; 50% below = *blue*; otherwise *orange*. Current dams are represented by *red circles*, planned dams by *green circles* (Cole et al., 2014).

the hydropower-dependent African countries and this includes DR Congo, Ghana, Tanzania, Egypt, Zambia, Mozambique, Malawi, and Nigeria.

The chapter also presented the river flow changes as well as current and planned hydropower dams (Fig. 2). Similar to Fig. 1, hydropower-dependent countries still dominate in terms of planned construction of hydropower dams. Eastern Africa dominates in the number of planned hydropower sites. Even though there is a predicted increase in river flow changes over areas around Chad and Niger, there is little indication of future hydropower developments. There is also a predicted reduction in river flow changes across most parts of Northern Africa, and this certainly does not favor future development of hydropower projects. Basins with expected increase in river flow are prominent in Eastern and certain portions of Western, Central, and Southern Africa. These areas provide suitable grounds for future consideration of sites for hydropower projects subject to other feasibility factors.

3.   CLIMATE CHANGE AND ENERGY PRODUCTION IN GHANA

3.1   Country Profile and Electric Power Production

Ghana is a West African country located between latitudes $4°44'$ N and $11°15'$ N and longitudes $3°15'$ W and $1°12'$ E. The country is bounded to the East by Togo, to the north by Burkina Faso, to the west by Ivory Coast, and to the south is the Gulf of Guinea. Endowed with several natural resources, the country covers an area of approximately 238,539 km^2, with a coastline spanning about 550 km (Armah and Amlalo, 1998). According to Ly (1980), this coastal area is split into three zones: the western, central, and eastern zones, mainly based on their geomorphologies. The Eastern coastal zone is about 149 km stretching from Aflao to the Laloi lagoon, west of Prampram. This zone is made up of barrier beaches, which consist of medium-to-coarse sand sediments. The Central Coastal Zone is a 321-km stretch of shoreline extending from Laloi Lagoon to the estuary of River Ankobra, consisting of rocky headlands and sand bars enclosing coastal lagoons. The Western Coastal Zone covers about 95 km of shoreline and consists of a fairly extensive beach and a coastal lagoon. This coastal zone extends from the estuary of the Ankobra River to the border with Cote d'Ivoire (Fig. 3).

According to the Institute of Statistical, Social and Economic Research (ISSER, 2005), Ghana's current electricity system has evolved through three distinct stages: the *Pre-Akosombo Era, the Hydroelectricity Era*, and the *Thermal Complementation Era*. In the *Pre-Akosombo Era* (i.e., before 1966), diesel generators were used to supply electricity to run Ghana's railway system as well as supply electric power to other cities including Accra, Koforidua, and Cape Coast. The power generated was not only expensive but also insufficient to expedite and maintain Ghana's industrialization program after independence (Owusu, 2010; Keener and Banerjee, 2005).

In response to the shortfall in power supply, the *Hydroelectricity Era* was conceived primarily to supply electricity to the Volta Aluminum Company (VALCO), to exploit Ghana's bauxite reserves. The Akosombo dam was then constructed to provide industries, commercial, and household consumers with a reliable and effective source of energy (ISSER, 2005). According to Turkson and Amadu (1999), as of 1972, the Akosombo dam could generate approximately 912 MW of electricity. The subsequent construction of the Kpong dam increased the total installed capacity to 1072 MW by the end of 1975 (Turkson and Amadu, 1999). In its entirety, the Volta River Authority's total installed capacity is 1930 MW (equivalent to about 8957 GWh). The Ghana Grid Company (GridCo), however, estimated a potential power deficit of 316 GWh (Energy Commission, 2010).

An additional 400-MW facility designed to generate about 1000 GWh of electricity has been constructed at Bui in the Brong Ahafo Region (i.e., the Bui Dam). At the moment, however, only about 90 MW of electricity is being produced (Boateng, 2014). There is also a construction of a 111-km gas pipeline from Atuabo to Aboadze (Boadu, 2014) to aid in the national electrification program. As at December 2015, the total installed energy capacity from the available energy generation sources in Ghana is 3656 MW as shown in Table 1. While Table 1 shows the contribution of individual energy sources to the total energy mix, Table 2 reveals the variation of the total energy generated in the past decade.

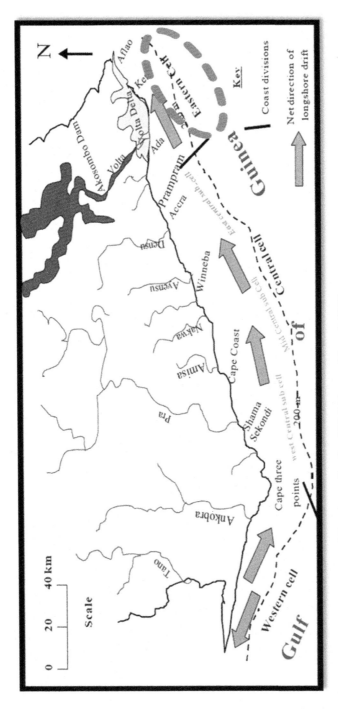

FIG. 3 Drainage and partitions of Ghana's coastline (Ly, 1980; Dickson and Benneh, 1988).

TABLE 1 Installed Electricity Generation Capacity as at the End of December 2015

	Plant	Fuel Type	Installed Capacity (MW)	Share (%)
Hydro	Akosombo	Water	1020	27.9
	Bui	Water	400	10.9
	Kpong	Water	160	4.4
	Subtotal		1580	43.2
Thermal	Takoradi Power Company (TAPCO)	LCO/NG	330	9.0
	Takoradi International Company (TICO)	LCO/NG	330	9.0
	Sunon Asogli Power (Ghana) Limited (SAPP)—IPP	NG	200	5.5
	Cenit Energy Ltd (CEL)—IPP	LCO	126	3.4
	Tema Thermal 1 Power Plant (TT1PP)	LCO/NG	110	3.0
	Tema Thermal 2 Power Plant (TT2PP)	DFO/NG	50	1.4
	Takoradi T3	LCO/NG	132	3.6
	Mines Reserve Plant (MRP)	DFO/NG	80	2.2
	Kpone Thermal Power Plant (KTPP)	NG	220	6.0
	Karpowership	HFO	225	6.2
	Ameri Plant	NG	250	6.8
	Subtotal		2053	56.2
Renewables	VRA Solar	Solar	2.5	0.1
	BXC Company	Solar	20	0.5
	Subtotal		22.5	0.6
Total			3656	100

DFO, diesel fuel oil; HFO, heavy fuel oil; LCO, light crude oil; NG, natural gas.
From Ghana Energy Commission.

TABLE 2 Electricity Generation by Plant (GWh) Per Installed Capacity (MW) in Ghana From 2006 to 2015

Plant	2006	2007	2008	2009	2010	2011	2012	2013	2014	2015
Hydro										
Akosombo	4690	3104	5254	5842	5961	6495	6950	6727	6509	4156
Kpong	929	623	941	1035	1035	1066	1121	1144	1148	819
Bui	–	–	–	–	–	–	–	362	730	870
Subtotal	5619	3727	6195	6877	6996	7561	8071	8233	8387	5845
Thermal										
TAPCO)	1416	1521	874	453	1234	1137	1061	1783	890	1784
(TICO)	1395	1417	1063	1040	1160	657	1168	1032	712	1336
(TT1PP)	–	–	–	570	591	559	622	475	697	541
(TRPP)	–	162	85	–	–	–	–	–	–	–
(ERPP)	–	80	45	–	–	–	–	–	–	–
(KRPP)	–	33	16	–	–	–	–	–	–	–
(MRP)	–	38	46	18	20	12	20	–	195	170
(TT2PP)	–	–	–	–	28	50	141	94	223	216
(SAPP)	–	–	–	–	138	1224	848	694	1255	1185
(CEL)	–	–	–	–	–	–	94	454	513	317
Takoradi T3	–	–	–	–	–	–	–	102	87	31

Continued

TABLE 2 Electricity Generation by Plant (GWh) Per Installed Capacity (MW) in Ghana From 2006 to 2015—cont'd

Plant	2006	2007	2008	2009	2010	2011	2012	2013	2014	2015
Karpowership	–	–	–	–	–	–	–	–	–	64
Ameri Plant	–	–	–	–	–	–	–	–	–	0
Subtotal	2811	3251	2129	2081	3171	3639	3953	4635	4572	5644
Renewables										
VRA solar								3	4	3
Total generation	8430	6978	8324	8958	10167	11200	12024	12870	12963	11492
Installed capacity	1730	1935	1,981	1970	2165	2170	2280	2831	2831	3656

CEL, Cenit Energy Ltd.; *ERPP*, Emergency Reserve Power Plant; *KRPP*, Kumasi Reserve Power Plant; *MRP*, Mines Reserve Plant; *SAPP*, Sunon Asogli Power (Ghana) Ltd.; *TAPCO*, Takoradi Power Company; *TICO*, Takoradi International Company; *TT1PP*, Tema Thermal 1 Power Plant; *TRPP*, Tema Reserve Power Plant; *TT2PP*, Tema Thermal 2 Power Plant.
Source: Ghana Energy Commission.

3.2   Climate Change Impacts

Global sea level has been projected to rise up to about 60 cm by 2100 in response to the combined processes of ocean warming and glacial melting (Bindoff et al., 2007). However, according to recent research, identified and accelerated decline of polar ice sheet mass increases the probability of future sea-level rise (SLR) of 1 m or more by 2100 (Allison et al., 2009; Rignot et al., 2008; Velicogna, 2009; Pfeffer et al., 2008; Lowe et al., 2009; Nicholls and Cazenave, 2010). According to Parry et al. (2007), toward the end of the 21st century, projected sea-level rise will affect low-lying coastal areas with large populations, and the cost of adaptation could amount to at least 5%–10% of Gross Domestic Product (GDP). They further state that new studies confirm that Africa is one of the most vulnerable continents to climate variability and change because of multiple stresses and low adaptive capacity.

Studies show that the sea-level rise in Ghana, which is in conformity with global trends, is estimated at about 2 mm/year (Ibe and Quelennec, 1989; Armah et al., 2005; Addo et al., 2008; Oteng-Ababio et al., 2011). However, with the continuous decline of ice sheets and glaciers through global warming, this estimate is likely to increase largely affecting not only the coastal zones but also the sea bed because as sea level rises, material on sandy shorelines is eroded from the upper beach and deposited on the near-shore ocean bottom (Bruun, 1962).

The adverse impacts of climate change on hydrological resources (particularly those used for electric power generation) could inevitably disrupt electrification purposes. In preparation for this distasteful occurrence, it is prudent to put alternative contingencies in place so as to maintain adequate electrification, should the anticipated impacts materialize.

3.3   Alternative Energy Sources—Renewable Energy

In order to largely meet the global energy demand and curtail the emission of CO_2 from fossil fuel sources, as well as in part, prepare for the impacts of climate change (i.e., on hydrological resources for electrification), alternative sources of energy production systems ought to be considered. Over the years, a few renewable energy sources—hydropower, solar, biogas, geothermal, wind, nuclear, and marine energy (Herzog et al., 2001)—have been identified and are being implemented worldwide. Within this cohort of renewables, the most promising sources (i.e., solar, wind, nuclear, small hydropower, and *marine energy*) are envisaged to play essential roles in meeting long-term global and local energy demands. The global energy contribution from these promising renewables, according to Herzog et al. (2001), could rise from 2% to about 20%–50% in the second half of the 21st century, with the provision of conducive measures.

The need to diversify energy sources addresses more than just the issue of global warming and climate change. Besides the sustenance of the natural environment and the quality of life of upcoming generations, it also secures the relative stability of the assortment of energy resources over the long term.

A major challenge, however, threatening the full adoption of renewables is that its efficiency is location specific. A particular renewable energy source that is adequate in one geographical region could be completely counterproductive if applied elsewhere without proper appraisals. For every geographical setting, therefore, the selection of particular renewable energy sources is governed by the physical (i.e., technical) and ancillary (i.e., environmental, social, political, and economic) constraints characteristic to the related resource. As a result, each energy resource has specific and restricted areas where their benefits can be optimally tapped. Despite the numerous sources of renewables, the scope of this chapter is limited to marine energy.

3.3.1   Marine Energy

Through the combined processes of Earth's rotation, temperature variation, wind, salinity and bathymetry, ocean processes are generated. Marine energy comprises ocean wave energy (8000–80,000 TWh per year), thermal gradient (~10,000 TWh per year), salinity gradient (~2000 TWh per year), ocean currents (>800 TWh per year), and tidal streams (>300 TWh per year) (Bard, 2007). The United Kingdom has already taken a bold step in adopting this technique of energy generation. With an average wave power exceeding 50 kW/m of wave front, the United Kingdom (particularly areas situated off the Scottish north and west coast) possesses substantial marine energy resources. In Scotland, for instance, about 45 TWh/year of electricity can be generated from offshore wave power estimated at 14 GW.

Usually, ocean currents maintain a nearly constant direction with marginal fluctuations in current speed, and this makes it suitable for installing energy extraction devices. Ocean current turbines can thus be used to harvest substantial marine energy when installed appropriately in areas with sufficient marine currents.

Ocean waves are generated via the transfer of wind energy. The size of the generated wave is mainly contingent upon the strength and duration of the prevailing wind and the available fetch. The ocean is a storehouse of great amounts of potential and kinetic energies. Unlike other forms of renewable energy such as wind and solar, the energy in waves is more concentrated (Drew et al., 2009). While the potential global power of ocean waves along the coasts is estimated at 1 TW (i.e., 10^{12} W), the wave power engendered in open seas is about 10^{13} W. This is largely due to the absence of energy losses due to friction and wave breaking (Panicker, 1976). Although both waves and wind are driven by the sun, the average power flow intensity of a wave is approximately ten times greater than in the sun's rays (Drew et al., 2009). This means that a device that can efficiently extract energy from waves can extract more energy per unit time than a device that harnesses solar or wind energy.

Due to the size and abundant energy density stored in the ocean, scientists and inventors have been inspired to design a number of devices for tapping this energy for human needs (Stahl, 1892; Leishman and Scobie, 1978; Shaw, 1982). Currently there are over 1000 patents covering *Wave Energy Conversion*

Devices (WECDs) (Falnes and Løvseth, 1991). WECDs are mechanical devices that are employed in harvesting the kinetic energy of ocean surface waves so as to generate electric power. Due to the variable nature of oceans, it's very common to design a particular WECD, which is unique to the geography and ocean state of the area. This is because the amount of energy harvested from ocean waves is a function of the amount of energy and power inherent in the ocean waves. The energy and the power in return are dependent on the wave height and the wave period. The energy (*E*) per unit wavelength in the direction of the wave, per unit width of wave front, and the power (*P*) per unit width of a wave front are given by the following equations:

$$E = \frac{\rho g^2}{\pi}\left(H^2 T^2\right) \tag{1}$$

$$P = \frac{1}{64}\frac{\rho g^2}{\pi}\left(H_s^2 T\right) \tag{2}$$

where ρ denotes the density of sea water (kg/m^3), g is the acceleration due to gravity (ms^{-2}), H is the wave height (m), T is the wave period (s), and *Hs* is the significant wave height (m) (Twidell and Weir, 1986). For further details of wave energy and the mathematical expressions, interested readers may refer to (Mollison, 1986; Thorpe, 1999; Duckers, 2004).

At the moment, a few WECDs have been implemented in some countries—the LIMPET Oscillating Water Column (Webb et al., 2005), Pelamis (Power, 2010), Salter's Duck (Greenhow et al., 2006), Aquamarine Power Oyster (Whittaker et al., 2007), and the Wave Dragon (Kofoed et al., 2006). Wave energy conversion devices, particularly the Pelamis, have portrayed vast strides in energy production. There are numerous advantages obtained from the use of WECDs. Firstly, wave energy extractors have a relatively minimal environmental footprint, even with respect to other sources of renewable energy. In particular, wave energy devices do not produce liquid, solid, or gaseous emissions (Thorpe, 1999). In addition, they are virtually carbon-free sources of energy production (Margheritini et al., 2012; Parkinson, 2013; Witt et al., 2012). Furthermore, waves travel long distances with minimal energy loss, meaning the installation of wave energy extractors is flexible (Drew et al., 2009). While wind turbines and solar panels must be placed in optimal locations, wave energy extractors can be placed in a larger area and still extract energy efficiently from the environment. Finally, wave energy extractors can generate electricity for approximately 90% of the day on average, compared to 20%–30% for wind turbines and solar panels (Drew et al., 2009). Therefore, wave energy extractors have the potential to be a more consistent source of energy.

3.3.2   *Impacts of Climate Change on Marine Energy*

While the observed discrepancy between general circulation models (GCMs) and their physical demonstrations signifies the alteration of wind patterns

(Hulme et al., 2002), a substantial body of literature shows that global wave heights have been changing over recent decades (Carter and Draper, 1988; Bacon and Carter, 1991). Although there is yet to be conclusive confirmation of its basis, global warming is attributed to the source of this change (Grevemeyer et al., 2000). Since ocean waves are propagated by wind action, significant variations in wind patterns will obviously impact the wave regimes, consequently translating into the potential wave energy that can be harvested. Relating wind speeds to wave power, a 5% change in wind speed will engender about 25% change in the wave power (Harrison and Wallace, 2005b).

The impacts of climate change on marine energy are more apparent since energy-converting systems are designed purposely to harvest the energy inherent in the oceans. A vivid example is the use of ocean wave energy conversion devices that aim at harvesting energies embedded in ocean waves owing to specific wave heights, periods, and directional ranges. Significant changes in met-ocean parameters will certainly affect the energy capture: Low-wind energies will propagate waves with low energies (Fig. 4).

On the other hand, high wind energy could bring about two scenarios: increased wave energy as a result of large waves will signify higher prospects of electric generation and/or potential threat and damage to WECDs (Wang et al., 2004). Some particular ocean current and tidal stream systems may be affected by amplified wave actions as a result of variation in ocean circulation (Greenhow et al., 2006). While devices fixed to the shoreline could be affected by a significant rise in sea levels, marine energy devices anchored in deeper waters could experience limited impacts, as a reduced amount of energy will be dissipated (Harrison and Wallace, 2005a).

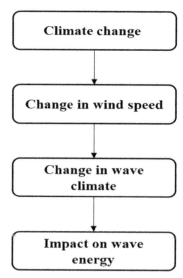

FIG. 4 Nexus between climate change and wave energy (Harrison and Wallace, 2005b).

4. MARINE ENERGY CONVERSION SYSTEMS

According to a report from the International Solar Energy Experts Workshop (Ahiataku-Togobo, n.d.), the National Renewable Energy Act stipulates a 10% target for renewable energy by the year 2020. On this premise, the onus lies on the government to approve of the private sector's involvement in attaining this target. Exploration into alternative sources of renewable and sustainable energy—solar, wind, and marine energy—is a step in the right direction if the government really wants to cut down the electric power deficit and also attain the 10% target. Each alternative has its own advantages as well as its short-comings. Nevertheless, among all the renewables, Mccullen et al. (2002) have shown that ocean waves contain the highest energy density. In addition, Pelc and Fujita (2002) have shown that the energy output from WECD far exceeds that of Solar and Wind; WECDs can produce power up to 90% of the time, compared to the approximately 20%–30% for wind and solar energy devices. Having the Gulf of Guinea and the areal expanse of the Atlantic Ocean at her disposal, it'll be prudent to research into this source of energy generation.

Drew et al. (2009) have shown that throughout the operational lifespan, WECDs require very low operational and maintenance costs. Depending on the type of WECD, there's little to no emissions during operation. The main parameters needed to facilitate a successful implementation of WECDs are the wave height and wave period, and these parameters can be easily assessed and predicted. The main concern, however, with WECDs is with the cost involved. As of now, it is difficult to assess the various impacts generated by the installa-tion of WECD in the ocean environment. That notwithstanding, it is well known that the proper orientation of the conversion system induces little environmental impacts and may also act as artificial reefs, increasing the quantity of marine species in the area. In the long term, as the patronage of renewables in the country surges, foreign investors are likely to invest. On the economic front, turning the focus to renewables will see the creation of numerous jobs that will help stabilize the nation's economy. Industrialization is also likely to shoot up coupled with the formation of secondary businesses that will render services in the renewables industry.

The implementation of the proposed marine energy conversion system is not a measure that seeks to replace the conventional and traditional energy genera-tion plants in Ghana; instead, it seeks to supplement the amount of energy the country is generating. This move will among other benefits ease the pressure on the energy sector to meet the growing demand of energy nationwide.

As recorded in a literature, the largest wave energies are localized within the temperate regions (Drew et al., 2009). Moving away from this region, the energies tend to reduce. With this provision, WECDs are generally designed with regard to the location and type (Drew et al., 2009). Nevertheless, due to the coastal geomorphology, seabed topography and met-ocean conditions, a few areas outside the temperate zones are potential areas for harnessing the ocean waves' energy.

These WECDs can be sited onshore, near-shore, and/or offshore depending on a range of factors. Shoreline devices have good proximity to the utility network, and this affords regular and easy monitoring and maintenance. Devices in near-shore localities are often secured to the seabed, giving it a suitable stationary base against which the conversion device can operate effectively. The energy content in deep water waves is very high (Drew et al., 2009); for this reason, greater amounts of energy can be harvested when WECDs are installed in deep waters. However, with regard to construction, these offshore WECDs are more challenging and expensive to maintain. Yet, for floating devices in deep water, the economic gains are high (Korde, 2000; Callaghan, 2006).

4.1 Prototype (Initial Design) of a Novel Marine Energy Conversion System

This system is designed to harness energy from the mass movement of water from both *ocean currents* and the *wave motion* on the surface of the ocean. The energy extraction begins from the moment the blades of the conversion system intercept the ocean currents and/or the supporting hinges are set into motion by incoming waves (Fig. 5). While the ocean currents cause an initial spin of the blades, the waves move the supporting hinges up and down. The combined up-and-down motion of the supporting hinges and the slow rotational speed of intercepting blades propel the spin of an attached turbine, which transfers energy to a slow-rotating shaft that is connected to gears. For this particular system, a *wells' turbine* is employed due to its ability to constantly spin in one direction, which allows for optimal energy harvesting. This energy is then transferred to drive a low-speed shaft that is also connected to gears. These gears alter the speed of the low-spinning shaft by increasing the spin via a high-speed shaft connected to a high-speed gear. The swift-spinning shaft drives the generator to produce electricity. Electricity from the generator will go to a transformer (i.e., a step-up transformer that will step up the voltage) via a transmission cable of the required kilo-volt (kV) rating.

The electricity, after being converted to the right voltage, is transmitted to an off-grid system. The transmission and distribution to end users then starts. The power take-off system is the turbine, which is directly attached to the generator (Fig. 5). The assignment of machinery and know-how from specific engineering industries will address the construction and installation issues.

The design of the conversion system takes into consideration the prevalence of coastal erosion in the area. The conversion system is thus fixed to a robust breakwater. The entire setup has been displayed in Fig. 5.

The energy extraction, particularly the marine current extraction, is slightly comparable to that of wind, which has received comprehensive attention from authors such as Ackermann, 2000; Tapia et al., 2001, 2004. The system has been designed taking into account the power coefficient (Cp), tip speed ratio (TSR) of the intercepting blades, and the thrust coefficient (Ct) (Eqs. 3–5) in addition

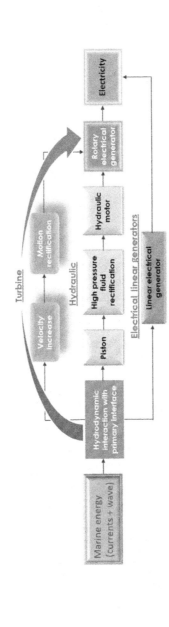

FIG. 5 Illustration of the power take-off mechanism—the turbine. *From Drew, B., Plummer, A.R., Sahinkaya, M.N., 2009. A review of wave energy converter technology. J. Power Energy 223(8), 887–902.*

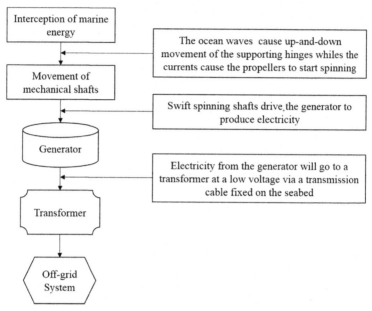

FIG. 6 Energy generation by the system.

to the met-ocean conditions that have a propensity of influencing the lifespan as well as the mechanical and operational efficiency of the entire system.

$$TSR = \frac{\omega R}{\upsilon_0} \tag{3}$$

$$C_t = \frac{T}{0.5\rho V_0^2 A} \tag{4}$$

$$C_p = \frac{P}{P_0} = \frac{P}{0.5\rho V_0^3 A} \tag{5}$$

where R denotes the radius of the turbine, ω is the rotational speed of the blade tip, ρ is the density of the sea water, A is the cross-sectional area of the rotor, and P is the power developed by the generator (Fig. 6).

4.2 Compatibility of the System in Ghana

Ghana has great potential with regard to the implementation of this marine energy conversion system. This opportunity is, however, elusive due to the dearth of substantial research in the field.

4.2.1 Geomorphological and Met-Ocean Feasibility

Given the location of Ghana, the prevalent winds will aid in the successful implementation of this device. The prevalent south-westerly winds travel from the Atlantic onto the continent. These winds are of great magnitude with seasonal variations and transmit their energy into the ocean. Ly (1980) and Boateng (2009) have shown that the prevailing south-westerly winds cause an oblique wave approach to the shoreline. This wave approach generates an east-ward littoral transport, which contributes to the erosion and accretion in the Eastern Coastal Zone. For this reason, the Eastern Coastal Zone is considered a high-energy beach (Ly, 1980). The nature of waves in this area demonstrates the possibility of installing the conversion system: The average wave height for the identified area is approximately 1.39 m with a mean period of about 10.91 s (Boatemaa et al., 2013). The issue of littoral transport will be addressed by the proper orientation and installation of the entire setup. Since calming of waves can be achieved by the installation of this system, coastal erosion in the area is likely to be significantly curtailed.

In addition, the sea bed topography of the estuary, from the continental shelf all the way into deep waters, consists of numerous canyons Manu et al. (2005). Waves that reach this point normally behave as though they were in deep waters, and this attribute aids in the propagation of relatively large waves that can be tapped by the WECD. Within this particular zone, the average depth of the coastal waters, the topography of the seabed, and the relatively low traffic density presuppose that implementation of the conversion system is practically achievable. In the event of SLR, the ensuing conditions are likely to favor the installation of the system.

When installed accurately, the device will reduce several environmental issues (Thorpe, 1999). The distribution of the generated energy will be done through an off-grid system. By and large, this is going to reduce the electric power deficit in the country. The advantages reaped from this project ranges from the availability of reliable electric power, which would inevitably be trans-lated into the decreased dependence and pressure on the presently available sources of energy production. This will cause a parallel increase in productivity due to the availability of uninterrupted power supply. In terms of the coastal stability of the Eastern Coastal Zone, the engineering measures put in place will curtail the rate of coastal erosion due to the dampening of waves (Shaw, 1982) by the system, and enhance a healthy rate of accretion. The system will act as a protection between the vulnerable coastline and the ocean, and insulate the coastal communities from coastal erosion and subsequent flooding. The most remarkable aspect of this project is that the source of the energy (i.e., the ocean) is free, unlimited, and at Ghana's disposal.

4.2.2 Economic Feasibility

Wave energy designers have no clear economic models to refer to, largely due to the embryonic stage of commercial WECDs (Vining and Muetze, 2009). Despite this, however, an attempt was made to conduct a brief economic assess-ment, peculiar to Ghana.

The cost of electricity (COE) is a critical standard by which most renewables are measured (Thorpe, 1999). The factors that control this cost include the cost of materials and the initial cost (IC). The size of the marine conversion system and the dependence on cable size are also a factor. For economic realizations, a cable size with the essential kilovolt rating and capacity to handle the load will be used to serve both the WECD and the ocean current conversion system. Further analyses would be carried out to establish a competitive FIT with regard to the energy generated from this system. This would be done by using the current cost of electricity in the country as a guide as well as precepts from countries like Ireland, Germany, Spain, and Portugal (ADENE, 2010). With advances in technology and an anticipated introduction of feed-in-tariffs (FiT) for wave energy, the cost of construction and consequently cost of electricity generated is expected to reduce significantly.

Currently, electricity generated from wave energy is $0.075 kWh in the United Kingdom. This price cannot be directly compared to electricity prices in Ghana ($0.08 kWh^{-1} for the lowest residential consumption, 0–50 kWh) considering the fact that technical expertise and machine components must be imported, which will invariably increase the cost of construction and maintenance of the facility. Insurance, taxes, and interest on loans linked to the operation of such a facility are also factors that will significantly influence the cost of electricity from wave energy (Fig. 7).

Financial institutions in Ghana, and other sub-Saharan African countries, still perceive the renewable energy projects (even solar and wind) as high-risk ventures, thereby serving as deterrents for both local and foreign investors (Boyd, 2012; Ajayi and Ajayi, 2013). Additionally, policy makers in Ghana are yet to make adequate provisions for the utilization of wave energy in Ghana. At the moment, the FiT in Ghana does not take into account electricity generation from wave energy. This incentive, which is usually linked to others such as tax exemptions and import duty waivers for renewable energy technology components, further increases the uncertainty (if not the cost of construction and operation) about the feasibility of a wave energy project. This situation further accentuates the need for R&D into WECD in the subregion as access to energy is vital in creating a dent in the existing energy poverty.

A 500-MW onshore facility according to ECO Northwest (2009) is estimated at approximately $750,000 per megawatt in order to generate energy at a competitive price to wind. This estimate, however, is without the inclusion of probable green energy subsidies or premium charges that wave energy may attract in the future. Table 3 shows the cost breakdown of an onshore wave energy farm as at 2008.

For a pilot phase, a conversion plant of adequate capacity may be appropriate to study the energy output as well as the environmental and the socioeconomic impacts. In this regard, the government could call on banks and other big companies in the nation so as to pool funds together to facilitate extensive research (AEA Energy and Sustainable Energy Ireland, 2006), and subsequent implementation of the marine energy conversion systems, as well as other sources of renewable and sustainable energy production. This of course will come with

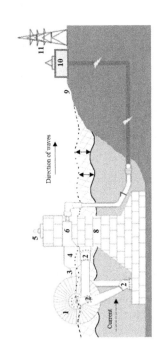

Shafts/ gears compartment + generator compartment connected to the transformer compartment:

Spinning of the shafts, controlled by the different gears initiates the rotation of the turbine that generates electric power in the generator. This electric energy is stepped up in the transformer compartment for transmission.

Interception blades/propellers:

Starts the spinning process upon interaction with the ocean currents.

Supporting hinges:

Moves up and down in response to the motion of incident waves

Operational lights:

These lights turn red when the device is in operation.

It serves as a safety precaution for marine vessels around the vicinity.

Sea-floor transmission cable:

Transmission of the generated electric power is achieved via this sea-floor cable of the required kV rating.

The distance from the propellers to the surface of the breakwater is sufficiently large to minimize the effect of reflection from the surface of the breakwater. The surface of the breakwater is further equipped with materials that absorb the wave energy, significantly reducing the amount of wave reflection.

1. Intercepting blades
2. Supporting hinges
3. Shafts/Gears compartment
4. Generator compartment
5. Operational lights
6. Transformer compartment
7. Sea-floor transmission cable
8. Offshore breakwater
9. Coast
10. Off-grid system
11. Overhead transmission lines

FIG. 7 Layout and description of the prototype.

TABLE 3 Construction Costs for 500-MW Wave Farm, 2008 (ECO Northwest, 2009)

Cost Category	Amount
Equipment: Onshore transformers and grid	$8,283,500
Equipment: Cables	$6,130,000
Equipment: Mooring	$38,089,000
Equipment: Power conversion modules	$202,701,000
Concrete structures	$77,878,000
Building/facilities	$21,379,000
Installation work	$19,973,000
Total	$374,433,500

some enticing incentives and subsidies for the banks and companies involved in this endeavor. Further, a portion of the funds allocated for the national energy development can be invested into renewables; hence, this project. Part of the proceeds from the Jubilee Oil Field could be channeled into the development of renewables upon parliamentary approval.

Privatization of power-generating companies that will produce off-grid power to certain parts of the nation must be seriously considered in the quest for energy independence. These private companies can distribute power to rural communities, reducing the pressure on ECG and other energy-distributing companies in the country.

It is the confidence of the authors that at a practical scale—when fully developed—the system would engender substantial energy and reliable power performance matrix. The technologies available as of now may seem cost intensive because most are in their budding stages; nevertheless, pursuing this endeavor is a worthwhile investment in the long term. By further optimizing and streamlining the conversion system, its operation and maintenance could incur little cost and generate more revenue for investors.

4.2.3 Challenges

Similar to other projects, the adoption of this system has some identified limitations. The transportation of the generated energy via seabed cables is likely to impact the marine species that reside on the seabed. Though there's no substantial research that quantifies the intensity of impact, the use of state-of-the-art materials will mitigate any grave impacts that may be caused in implementing this mode of energy generation. Shoreline and near-shore devices have a common hindrance in which shallow water leads to waves and currents with reduced power, limiting the harvesting potential of the system. With this problem, advanced engineering techniques would be further explored to fine tune and augment the harvesting prospects of the system. Corrosion is another issue of major concern (Mccullen et al., 2002), which would be addressed by the choice of adequate and noncorrosive coating materials.

5. CONCLUSION

Harnessing the energy from the ocean is a very efficient and green way of producing sustainable energy, by employing wave and ocean current energy conversion techniques. The ocean is a massive energy resource, and harnessing its energy signifies a vital step toward meeting renewable sustainable energy targets in the country. With the current energy crises in Ghana, it is imperative that further research is conducted on the implementation of this novel system for marine energy harvesting as well as other renewables that seek to mitigate the power shortfall. The gains of this project are obvious; the development of which is sustainable, as it combines crucial economic, environmental, global, and social factors.

The eastern coastal zone has been identified as a high-energy beach, presupposing that the implementation of the proposed system in the area is feasible. Nonetheless, like other renewables, the materialization of this conversion system into operational status will span a few years. This will involve further R&D—the conduction of series of tests and analyses as well as investigation of its development and challenges especially in the face of global climate change. The assistance and contribution by way of funding and other engineering expertise from engineers and researchers in related fields as well as decision and policy makers is being solicited to put this new system into motion.

In conclusion, it is obvious that pursuing this endeavor is worthwhile in the long term. The social, environmental, and economic aspects are all in the long term positively impacted. The best news is that the energy source of this marine conversion system is free and unlimited.

In a nutshell, there exists a bright prospect for Africa in terms of hydropower development, which includes marine energy sources for countries along the coast. However, plans for future hydropower sites are evident in countries that are already depending heavily on hydropower sources. This situation necessitates the need for greater advocacy and support for hydropower projects in Africa by all relevant stakeholders as the continent presses toward mitigating climate change impacts and improving assess to energy.

Acknowledgments

The authors will like to express their profound gratitude to Prof. Yongping Chen, Nii Okpoti Adjei Commey, Dr. Jingyu Huang, and Philip Nti-Nkrumah as well as colleagues and reviewers who made matchless input into this paper. Their assistance is gratefully acknowledged.

REFERENCES

Ackermann, T., 2000. Wind energy technology and current status: a review. Renew. Sustain. Energy Rev. 4 (4), 315–374.

Addo, K.A., Walkden, M., Mills, J.P., 2008. Detection, measurement and prediction of shoreline recession in Accra, Ghana. ISPRS J. Photogramm. Remote Sens. 63 (5), 543–558.

ADENE, 2010. Feed-in tariff for wind energy in Portugal. Agencia Para a Energia, Lisbon.

AEA Energy, Sustainable Energy Ireland, 2006. Review and Analysis of Ocean Energy Systems Development and Supporting Policies. A report by AEA Energy & Environment on the behalf of Sustainable Energy Ireland (SEI) for the IEA's Implementing Agreement on Ocean Energy Systems, p. AEA Energy & Environment, UK. Available at: http://www.iea-oceans.org/_fich/6/Review_Policies_on_OES_2.pdf. Ref.: ED023530014.

Ahiataku-Togobo, W., Ghana's renewable energy act—Opportunities and prospects. Available at: http://energycenter.knust.edu.gh/enc/downloads/16/161786.pdfAccessed February 1, 2015.

Ajayi, O.O., Ajayi, O.O., 2013. Nigeria's energy policy: Inferences, analysis and legal ethics toward RE development. Energy Policy 60, 61–67.

Akella, A.K., Saini, R.P., Sharma, M.P., 2009. Social, economical and environmental impacts of renewable energy systems. Renew. Energy 34 (2), 390–396.

Allison, I., et al., 2009. Ice sheet mass balance and sea level. Antarct. Sci. 21 (05), 413–426.

Armah, A.K., Amlalo, D.S., 1998. Coastal zone profile of Ghana: Accra, Gulf of Guinea large marine ecosystem project. Ministry of Environment, Science and Technology, Accra.

Armah, A.K., et al., 2005. Sea-level rise and coastal biodiversity in West Africa: a case study from Ghana. In: Climate Change and Africa. Cambridge University Press, Cambridge, pp. 204–217. https://doi.org/10.1017/CBO9780511535864.029.

Arnell, N.W., 1999. Climate change and global water resources. Glob. Environ. Change 9, S31–S49.

Ayanda, M., 2011. Small and medium scale enterprises as a survival strategy for employment generation in Nigeria. J. Sustain. Dev. 4 (1), 200–206.

Bacon, S., Carter, D.J.T., 1991. Wave climate changes in the North Atlantic and North Sea. Int. J. Climatol. 11 (5), 545–558.

Bard, J., 2007. Ocean Energy Systems. Available at: http://www.webcitation.org/6ZhLw1yYn.

Bates, B. et al., 2008. Climate Change and Water: Technical Paper VI, Intergovernmental Panel on Climate Change (IPCC), Geneva.

Bindoff, N.L., et al., 2007. Observations: Oceanic Climate Change and Sea Level. Intergovernmental Panel on Climate Change (IPCC), Cambridge University Press, Cambridge.

Boadu, K.A., 2014. Atuabo gas processing plant: Solution to energy crisis?. Available at: http://graphic.com.gh/features/features/24155-atuabo-gas-processing-plant-solution-to-energy-crisis.htmlAccessed February 1, 2015.

Manson Awo Akosua Boatemaa, Apppeaning Addo Kwasi, Adelina Mensah, Impacts of shoreline morphological change and sea level rise on mangroves: the case of the keta coastal zone. J. Environ. Res. Manag. Vol. 4(11). pp. 0359-0367, 2013. Available at http://www.e3journals.org/cms/articles/1387371004_Kwasi et al.pdf [Accessed 30 January 2015].

Boateng, I., 2009. Development of integrated shoreline management planning: a case study of Keta, *Ghana:* Proceedings of the Federation of International Surveyors Working Week.

Boateng, K.A., 2014. De-lighting Ghanaians, the state of Ghana's power sector—citifmonline. Available at: http://www.citifmonline.com/2014/06/02/de-lighting-ghanaians-the-state-of-ghanas-power-sector/#Accessed February 1, 2015.

Boyd, A., 2012. Informing international UNFCCC technology mechanisms from the ground up: Using biogas technology in South Africa as a case study to evaluate the usefulness of potential elements of an international technology agreement in the UNFCCC negotiations process. Energy Policy 51, 301–311.

Bruun, P., 1962. Sea-level rise as a cause of shore erosion. J. Waterw. Harb. Div. 88 (1), 117–132.

Callaghan, J., 2006. Future Marine Energy Results of the Marine Energy Challenge: Cost Competitiveness and Growth of Wave and Tidal Stream Energy. Carbon Trust, London 40.

Callaghan, J., Boud, R., 2006. Future Marine Energy—Results of the Marine Energy Challenge: Cost Competitiveness and Growth of Wave and Tidal Stream Energy. Carbon Trust, London. Available from http://www.thecarbontrust.co.uk/carbontrust/about/publications/FutureMarineEnergy.pdf.

Carter, D.J.T., Draper, L., 1988. Has the north-east Atlantic become rougher? Nature 332, 494.

Chevallier, P., et al., 2011. Climate change threats to environment in the tropical Andes: glaciers and water resources. Reg. Environ. Change 11 (1), 179–187.

Christensen, N.S., et al., 2004. The effects of climate change on the hydrology and water resources of the Colorado River basin. Clim. Change 62 (1-3), 337–363.

Cole, M.A., Elliott, R.J.R., Strobl, E., 2014. Climate change, hydro-dependency, and the African Dam Boom. World Dev. 60, 84–98.

Crabtree, G.W., Dresselhaus, M.S., Buchanan, M.V., 2004. The Hydrogen Economy. Phys. Today 57 (12), 39.

Dickson, K.B., Benneh, G., 1988. A New Geography of Ghana. Longmans, London, UK.

Drew, B., Plummer, A.R., Sahinkaya, M.N., 2009. A review of wave energy converter technology. J. Power Energy 223 (8), 887–902.

Duckers, L., 2004. Wave energy. In: Renewable Energy: Power for a Sustainable Future. vol. 2. Oxford University Press, Oxford, UK, pp. 297–340.

ECO Northwest, 2009. Economic Impact Analysis of Wave Energy: Phase One, Available from https://ir.library.oregonstate.edu/xmlui/bitstream/handle/1957/15692/Economic-Impact-Analysis-of-Wave-Energy-Phase-One—Version%202.pdf?sequence=6.

Edwards, P.P., et al., 2008. Hydrogen and fuel cells: towards a sustainable energy future. Energy Policy 36 (12), 4356–4362.

Energy Commission, 2010. 2010 ENERGY (SUPPLY AND DEMAND) OUTLOOK FOR GHANA Draft. . Accra.

Falnes, J., Løvseth, J., 1991. Ocean wave energy. Energy Policy 19 (8), 768–775.

Flavin, C., O'Meara, M., 1997. Financing solar electricity: the off-the-grid revolution goes global. World Watch 10, 28–36.

Gaudard, L., Gilli, M., Romerio, F., 2013. Climate change impacts on hydropower management. Water Resour. Manag. 27 (15), 5143–5156.

Greenhow, M. et al., 2006. A theoretical and experimental study of the capsize of Salter's duck in extreme waves. J. Fluid Mech., 118(1), p. 221. Available at http://www.mendeley.com/research/theoretical-experimental-study-capsize-salters-duck-extreme-waves/ [Accessed 27 January 2015].

Grevemeyer, I., Herber, R., Essen, H.-H., 2000. Microseismological evidence for a changing wave climate in the northeast Atlantic Ocean. Nature 408 (6810), 349–352.

Hagemann, S., et al., 2013. Climate change impact on available water resources obtained using multiple global climate and hydrology models. Earth Syst. Dyn. 4, 129–144.

Hamududu, B., Killingtveit, A., 2012. Assessing climate change impacts on global hydropower. Energies 5 (2), 305–322.

Harrison, G.P., Wallace, A.R., 2005a. Climate sensitivity of marine energy. Renew. Energy 30 (12), 1801–1817.

Harrison, G.P., Wallace, A.R., 2005b. Sensitivity of wave energy to climate change. IEEE Trans. Energy Convers. 20 (4), 870–877.

Herzog, A.V., Lipman, T.E., Kammen, D.M., 2001. Renewable energy sources. In: Encyclopedia of Life Support Systems (EOLSS). EOLSS Publishers, Oxford, UK. Forerunner Volume-'Perspectives and Overview of Life Support Systems and Sustainable Development.

Hulme, M., et al., 2002. Climate Change Scenarios for the United Kingdom: the UKCIP02 Scientific Report, Tyndall. Centre for Climate Change Research, School of Environmental Sciences, University of East Anglia, Norwich.

Ibe, A.C., Quelennec, R.E., 1989. Methodology for Assessment and Control of Coastal Erosion in West and Central Africa. UNEP, Geneva. UNEP Regional Seas Reports and Studies no. 127.

IEA (International Energy Agency), 2013. Key world energy statistics. IEA (International Energy Agency), Paris.

IPCC, 2011. IPCC Special Report on Renewable Energy Sources and Climate Change Mitigation. Prepared By Working Group III of the Intergovernmental Panel on Climate Change. Cambridge University Press, Cambridge, UK.

IPCC (Intergovernmental Panel on Climate Change), 2007. Africa. Cambridge University Press, Cambridge.

IRENA, 2013. Africa's Renewable Future—The Path to Sustainable Growth. IRENA, Abu Dhabi.

ISSER, 2005. Guide to Electric Power in Ghana, first ed. University of Ghana, Accra.

Keener, S., Banerjee, S.G., 2005. Ghana: Poverty and Social Impact Analysis of Electricity Tariffs. ESMAP, Washington.

Kofoed, J.P., et al., 2006. Prototype testing of the wave energy converter wave dragon. Renew. Energy 31 (2), 181–189. Available at http://www.mendeley.com/catalog/prototype-testing-wave-energy-converter-wave-dragon/. [Accessed 27 January 2015].

Korde, U.A., 2000. Control system applications in wave energy conversion. In: Proceedings OCEANS 2000. IEEE. pp. 1817–1824.

Leishman, J.M., Scobie, G., 1978. The Development of Wave Power: A Techno-economic Study. Economic Assessment Unit, National Engineering Laboratory, Department of Industry, Glasgow, Scotland.

Lowe, J., et al., 2009. UK Climate Projections Science Report: Marine and Coastal Projections. Meteorological Office, Hadley Center, Exeter, UK.

Ly, C.K., 1980. The role of the Akosombo Dam on the Volta River in causing coastal erosion in central and eastern Ghana (West Africa). Mar. Geol. 37 (3), 323–332.

Manu, T., Botchway, I.A., Apaalse, L.A., 2005. Petroleum exploration opportunities in the Keta Area. Volta Monitor 1, 56–57.

Margheritini, L., Hansen, A.M., Frigaard, P., 2012. A method for EIA scoping of wave energy converters—based on classification of the used technology. Environ. Impact Assess. Rev. 32 (1), 33–44. Available at http://www.mendeley.com/catalog/method-eia-scoping-wave-energy-convertersbased-classification-used-technology/. [Accessed 27 January 2015].

Markoff, M.S., Cullen, A.C., 2008. Impact of climate change on Pacific Northwest hydropower. Clim. Change 87 (3–4), 451–469.

Mccullen, P., et al., 2002. Wave energy in Europe: current status and perspectives. Renew. Sustain. Energy Rev. 6, 405–431.

McDowall, W., Eames, M., 2006. Forecasts, scenarios, visions, backcasts and roadmaps to the hydrogen economy: a review of the hydrogen futures literature. Energy Policy 34 (11), 1236–1250.

Mensah, S., 2004. In: A Review of SME Financing Schemes in Ghana. A Presentation at the UNIDO Regional Workshop of Financing SMEs, Accra, Ghana in March 2004.

Mideksa, T.K., Kallbekken, S., 2010. The impact of climate change on the electricity market: a review. Energy Policy 38 (7), 3579–3585.

Mollison, D., 1986. Wave climate and the wave power resource. In: Hydrodynamics of Ocean Wave-Energy Utilization. Springer, Berlin, Heidelberg, pp. 133–156.

Mundial, B., 1996. Meeting the challenge for rural energy and development. In: Documento de trabajo. Banco Mundial, Washington, DC.

Nicholls, R.J., Cazenave, A., 2010. Sea-level rise and its impact on coastal zones. Science 328 (5985), 1517–1520.

Oteng-Ababio, M., Owusu, K., Addo, K.A., 2011. The vulnerable state of the Ghana coast: The case of Faana-Bortianor. Jàmbá: J. Disaster Risk Stud. 3 (2), 429–442.

Owusu, A., 2010. Towards a Reliable and Sustainable Source of Electricity for Micro and Small Light Industries in the Kumasi Metropolis. Kwame Nkrumah University of Science and Technology, Kumasi, Ghana.

Painuly, J.P., 2001. Barriers to renewable energy penetration; a framework for analysis. Renew. Energy 24 (1), 73–89.

Panicker, N.N., 1976. In: Power resource potential of ocean surface waves. Proceedings of the Wave and Salinity Gradient Workshop, Newark, Delaware, USA. pp. J1–J48.

Parkinson, G., 2013. Oceanlinx Launches World's First 1 MW Wave Energy Machine In South Australia. Clean Technologies. Available at: http://www.mendeley.com/research/oceanlinx-launches-worlds-first-1-mw-wave-energy-machine-south-australia/ [Accessed 27 January 2015].

Parry, M.L., et al., 2007. Climate Change 2007: Impacts, Adaptation and Vulnerability. In: Contribution of Working Group II to the Fourth Assessment Report of the Intergovernmental Panel on Climate Change. Cambridge University Press, Cambridge, UK.

Pelc, R., Fujita, R.M., 2002. Renewable energy from the ocean. Mar. Policy 26 (6), 471–479. Available at http://www.mendeley.com/catalog/renewable-energy-ocean-1/. [Accessed 12 December 2014].

Pfeffer, W.T., Harper, J.T., O'Neel, S., 2008. Kinematic constraints on glacier contributions to 21st-century sea-level rise. Science 321 (5894), 1340–1343.

Piao, S., et al., 2010. The impacts of climate change on water resources and agriculture in China. Nature 467 (7311), 43–51.

Power, P.W., 2010. Pelamis Wave Power. Director. Available at http://www.mendeley.com/catalog/pelamis-wave-power/. [Accessed 27 January 2015].

Rignot, E., et al., 2008. Recent Antarctic ice mass loss from radar interferometry and regional climate modelling. Nat. Geosci. 1 (2), 106–110.

Scanlon, T.M., et al., 2007. Positive feedbacks promote power-law clustering of Kalahari vegetation. Nature 449 (7159), 209–212.

Shaw, R., 1982. Wave Energy: A Design Challenge. E. Horwood.

Solomon, S., 2007. Climate Change 2007—The Physical Science Basis: Working Group I Contribution to the Fourth Assessment Report of the IPCC. Cambridge University Press, Cambridge, UK.

Stahl, A., 1892. The utilization of the power of ocean waves. Trans. ASME 13, 438.

Tapia, A., Tapia, G., Ostolaza, J.X., 2004. Reactive power control of wind farms for voltage control applications. Renew. Energy 29 (3), 377–392. Available at http://www.mendeley.com/catalog/reactive-power-control-wind-farms-voltage-control-applications-3/. [Accessed 27 January 2015].

Tapia, A., et al., 2001. Reactive power control of a wind farm made up with doubly fed induction generators. I. In: 2001 IEEE Porto Power Tech Proceedings (Cat. No.01EX502). IEEE, pp. 6. Available at http://www.mendeley.com/research/reactive-power-control-wind-farm-made-up-doubly-fedinduction-generators-i/. [Accessed 27 January 2015].

Thorpe, T.W., 1999. A Brief Review of Wave Energy. Harwell Laboratory, UK.

Turkson, J.K., Amadu, M.B., 1999. Environmental protection implications of the electric power restructuring in Ghana. (Denmark. Forskningscenter Risoe. Risoe-R; No. 1138(EN)). (UNEP Collaborating Centre on Energy and Environment. Working Paper; No. 8). Available from: http://orbit.dtu.dk/files/7729716/ris_r_1138.pdf.

Twidell, J., Weir, T., 1986. Renewable Energy Resources. Taylor & Francis, Abingdon. Available at http://www.mendeley.com/catalog/renewable-energy-resources-33/. [Accessed 27 January 2015].

USAID, 1999. An energy roadmap for Ghana: From crisis to the fuel for "economic freedom". Available from http://pdf.usaid.gov/pdf_docs/Pnace859.pdf.

Velicogna, I., 2009. Increasing rates of ice mass loss from the Greenland and Antarctic ice sheets revealed by GRACE. Geophys. Res. Lett. 36 (19).

Vining, J.G., Muetze, A., 2009. Economic factors and incentives for ocean wave energy conversion. IEEE Trans. Ind. Appl. 45 (2), 547–554.

Vörösmarty, C.J., et al., 2000. Global water resources: vulnerability from climate change and population growth. Science 289 (5477), 284–288.

Wang, X.L., Zwiers, F.W., Swail, V.R., 2004. North Atlantic ocean wave climate change scenarios for the twenty-first century. J. Clim. 17 (12), 2368–2383.

Webb, I., Seaman, C., Jackson, G., 2005. Oscillating water column wave energy converter evaluation report. Carbon. Available at http://www.mendeley.com/catalog/oscillating-water-column-wave-energy-converter-evaluation-report-1/. [Accessed 27 January 2015].

Whittaker, T. et al., 2007. The development of Oyster—a shallow water surging wave energy converter. 7th European Wave and Tidal Energy Conference. Available at: http://www.mendeley.com/catalog/development-oyster-shallow-water-surging-wave-energy-converter/ [Accessed 27 January 2015].

Witt, M.J. et al., 2012. Assessing wave energy effects on biodiversity: the wave hub experience. Philos. Trans. A Math. Phys. Eng. Sci., 370(1959), pp. 502–29. Available at: http://www.mendeley.com/catalog/assessing-wave-energy-effects-biodiversity-wave-hub-experience/ [Accessed 23 December 2014].

World Bank, 2013. The Africa Competitiveness Report 2013. World Bank, Washington, DC.

Chapter 2

Promoting Research in Sustainable Energy in Africa—The CIRCLE Model

Benjamin Apraku Gyampoh
African Academy of Sciences (AAS), Nairobi, Kenya

Chapter Outline

1. Introduction	29	3. Driving Solutions Through	
2. The CIRCLE Approach	30	Intra-African Research	
2.1 Supporting Different		Collaboration	32
Needs of ECRs	30	4. CIRCLE-Funded Research	
2.2 Intra-African Collaboration	31	in Energy	33
2.3 Keeping the Links With the		5. Climate Change, Hydropower,	
North	31	and Energy in Africa	34
2.4 Creating a Conducive		6. Research Uptake and the	
Environment for Continuous		Hydropower Energy Conference	34
Researcher Development	31	7. Conclusion	35
2.5 Research Uptake	32	References	35

1. INTRODUCTION

Climate Impacts Research Capacity and Leadership Enhancement (CIRCLE) has become a very effective cross-cutting platform allowing researchers with different expertise and experience to converge, undertake studies, and network to find solutions to Africa's challenges that are being exacerbated by climate change. Research undertaken by CIRCLE researchers are under five thematic areas—Water, Agriculture, Health and livelihoods, Energy, and Policy.

CIRCLE approach focuses on human capital and institutions not as exclusive components for the researcher and research development but as joint components of one system. CIRCLE recognizes human capital as being the most cost-effective and viable entry point for sustaining both organizational and institutional research capacity in Africa. Available evidence suggests that increasing human capital has positive impacts on overall organizational and

Sustainable Hydropower in West Africa. https://doi.org/10.1016/B978-0-12-813016-2.00002-2
29

institutional performance (Garavan and McGuire, 2001; Seleim et al., 2007). Regarding institutions, CIRCLE emphasizes providing critical support for institutions to build on strengths of institutions and address weaknesses. This is done by first assisting the institutions to undertake and needs/gaps assessment to identify where support is needed. Identifying these and addressing them is to ensure that the researchers who come on the CIRCLE program as CIRCLE Visiting Fellows (CVFs) would return after their one-year fellowship to work in better enabling environments. The program is building a critical mass of Climate Change Impact researchers.

2. THE CIRCLE APPROACH

While many of the issues faced by researchers in pursuing an academic career and becoming an independent researcher are global, Africa's researchers and institutions face a host of competing pressures and challenges that make it even more difficult than their colleagues in the north. Early-Career Researchers (ECRs) based in Africa do not have adequate opportunities to deepen their scientific and technical abilities in a structured system once they complete a PhD program and get absorbed into a university or a research institute. The absence of protected time for research and research-funding opportunities in most African universities result in a block early in the pipeline of producing world-class researchers that Africa desperately needs for the development of knowledge-based economies and addressing the continent's challenges.

The CIRCLE model is designed to support the development of Africa's future research leaders in addressing the challenges and taking advantage of opportunities that climate change offers, and to accelerate the development of Africa's development. In addition to being provided with personal fellowships with research funding to undertake their studies, the program offers its CVFs intensive personal mentoring and benefit from a comprehensive program of capacity-building and networking and collaboration activities alongside their peers.

2.1 Supporting Different Needs of ECRs

CIRCLE offers two categories of fellowships for ECRs in African institutions. There is a Post-Master's category and a Post-PhD category. A post-Master's fellowship category is introduced into the program alongside the usual postdoc known because it is recognized that in Africa, there are lecturers or researchers in institutions who do not have a PhD yet and such a program would afford them a year of research work and concentration in preparing to undertake a PhD study.

The one-year fellowship where researchers undertake their research away from their home institutions brings many advantages to the researchers. One is exposure to new research systems and ideas in the institutions and countries the researchers relocate to. The researchers are expected to get an injection of new ideas and exposure to different studies and perspectives at where they are undertaking the fellowship. Another is time off normal workloads to think and create solutions for the continent. Most researchers and lecturers in Africa are responsible for the large number of students that they have to teach, administer tests, and mark scripts. The workload usually makes it impossible to devote more time to research. One year away from the normal schedules affords a great opportunity to concentrate on research.

2.2 Intra-African Collaboration

Intra-African collaboration is also considered very important in arriving at decisions that would be of relevance and applicable across countries while learning of best practices in institutions from different countries. Long-term relationships are promoted and facilitated to ensure the relationships are lasting and can develop into bigger programs in Africa. One of the outputs is the signing of a Memorandum of Understanding between the University of Ibadan, Nigeria, and the University of Energy and Natural Resources (UENR) in Ghana. This working relationship came about because of the fellowship of a researcher from UENR to the University of Ibadan. There is also building of networks among the researchers that is anticipated to lead to more collaboration between African researchers as these ECRs develop into the research leaders on the continent.

2.3 Keeping the Links With the North

The strong focus on intra-Africa collaboration was pursued alongside ensuring the relationship with researchers from the north is maintained. There are experts referred to in the program as "Specialist Advisors" who are carefully selected and well briefed to support the work that is being carried out in the African institutions by the CIRCLE Visiting Fellows. This relationship is structured to ensure there is a crossing of knowledge between the African researcher and the advisor from the northern institution. The specialist advisor is just as the name denotes. The person is an expert and provides advice. The Specialist Advisor is not to direct the work of the CVF. The CVF owns the work as the Principal Investigator (PI) with a supervisor at his host institutions and a mentor in his home institution.

2.4 Creating a Conducive Environment for Continuous Researcher Development

The fellowship program is one of two components of the CIRCLE program; the second is the Institutional Strengthening Program. Home institutions for the CIRCLE Visiting Fellows are supported to plan for strengthening their own research training and commissioning systems, revising their research strategies, and reviewing their staff career development programs. Each institution is assisted to do and institutional research needs/gaps assessment based on which suggested interventions can be made to improve the systems and structured at the institution for research. This is very critical to ensure the CVF returns after the fellowship to an institution with an improved system to support their research work and career development.

2.5 Research Uptake

Knowledge generated from CIRCLE must be used to offer change and better lives. It takes time and effort to get knowledge generated from research to reach the right people that require it to act. This adoption of knowledge is what CIRCLE facilitates through its research uptake component. Understanding adoption as a process helps to realistically influence it, and to know how to apply methods to help achieve it (Andrews, 2012). The studies undertaken by the CIRCLE researchers must be relevant and usable by the decision makers of the various countries. It should also be very useful to the communities in which the studies are undertaken. The interest of CIRCLE is that the funds put into the research should not only be about getting publications in scientific journals. The communities who provide the data that researchers utilize must feel part of the program and willing to contribute to the work of finding solutions to improve their adaptive capacity. A good way of getting such cooperation is to involve them right from the conception of the study, work with them, and go back to them to present the findings of the study as well as the recommendations to improve on how they do things.

3. DRIVING SOLUTIONS THROUGH INTRA-AFRICAN RESEARCH COLLABORATION

One thing the CIRCLE program places much premium on is "promoting intra-African research collaborations." Research collaboration of any form is important, especially in multidisciplinary areas such as climate change. When there is a high proportion of African scientists collaborating with researchers outside Africa, it is usually interpreted as a positive aspect in scientific knowledge production. However, when this collaboration is too high and there is very little of similar collaboration among African researchers based in Africa, it may just be a manifestation of dependence on foreign researchers or lack of confidence in collaborating with colleagues on the continent. Perhaps most of the funding for research coming from outside Africa and the related grant conditions may be driving the high number of African external collaborations. This is something the CIRCLE program took note of and devised means of addressing this lack of research collaboration within Africa. Scientific collaboration is the backbone of research and is critical to the establishment of the African Union (AU) as a world player (AOSTI, 2014). Collaboration between AU members occurred in only 4.1% of AU scientific papers in 2005–2007 and in 4.3% of the papers in 2008–2010 (AOSTI 2014). Through the placement of African researchers from one institution in Africa in another African institution for a year working under the supervision of a senior researcher, there have been many collaborations developed under the program. This is deepening more collaboration between African researchers.

4.   CIRCLE-FUNDED RESEARCH IN ENERGY

CIRCLE has defined five thematic areas for research done on the program. CIRCLE Visiting Fellows independently develop their proposals based on their identified problems and their fields of expertise. All CIRCLE-funded research have at least two of the five thematic areas covered. But broadly, the research has been clustered under the five thematic areas based on the dominant theme of the research. From the clustering, 8 out of the 100 successful research proposals that have been selected and funded by CIRCLE fall under the thematic area of energy. This represents just 8% of the total studies supported. This small proportion also indicates the status of energy and climate change research in Africa. Most studies in climate change are centered around agriculture. Energy is very important and demands conscious effort to increase knowledge on the impact of climate change on energy in Africa

Breakdown of CIRCLE-funded research by thematic areas

Thematic Area	Cohort 1	Cohort 2	Cohort 3	Total
Agriculture	18	14	11	43
Energy	2	2	4	8
Health and Livelihoods	6	8	8	22
Policy	5	4	9	18
Water	3	1	5	9
Total	34	29	37	100

The eight studies undertaken under the energy cluster of CIRCLE are:

i. Potential Impact of Climate Change on Hydropower Generation in Ghana, West Africa
ii. Low-carbon development strategies: issues and challenges in the West African Power Pool electricity
iii. Assessment and Analysis of Renewable Energy Technology for the Nigeria Coastline Communities
iv. Renewable Energy/Solar Transition Landscape in South Africa in the Face of Climate Change

v. Energy and Resource Recovery from Litters Generated in a Community for Reducing Greenhouse Gas Emission and Mitigating Climate Change Effects

vi. Climate Change Adaptation as it affects Pyrolysis of Lignocellulosic Materials from Oil Palm Fruit Fiber and Physic Nut Shell at Low Temperature and Pressure for Energy and Chemical Productions

vii. Smart Agroforestry-Based Biofuel Systems: Creating Integrated Food-Energy Systems in Kenya

viii. Contributions of Plastic Waste innovations to Greenhouse Gas Reductions: Role of Informal Sector

5. CLIMATE CHANGE, HYDROPOWER, AND ENERGY IN AFRICA

Generally, Africa is an energy-poor continent. Most Africans do not have access to energy and even those who have do not have regular and stable access to energy. There has been expansion in renewable energy use in recent years, except for hydropower (Quitzow et al., 2016). In West Africa, gas accounts for almost half of electricity generation, while oil and hydropower make up most of the remainder. Climate change and the resulting changes in precipitation and temperature regimes will affect hydropower generation. For much of West Africa, forecast on precipitation is not certain but that notwithstanding climate change is expected to be a major threat to Hydropower development in West Africa.

The uncertainty of projections and the need for more energy coupled with the weak and fragile economies of most countries in West Africa require extensive studies to find solutions to the challenges that climate change could pose for energy production, especially hydropower. What mitigation measures for hydropower generation and designs should be developed in response to the effects and impacts of climate change. This requires funding to support these studies.

The CIRCLE program identified the study "Potential Impact of Climate Change on Hydro-Power Generation in Ghana, West Africa" as one of such critical studies that the continent needs. Funding this study and the needed interest that it generated testifies to the importance and the need for the knowledge.

6. RESEARCH UPTAKE AND THE HYDROPOWER ENERGY CONFERENCE

To ensure that the work undertaken by CIRCLE Visiting Fellows do reach the right end users, the program created a research uptake fund where the researchers can access funding to send their research findings to the right audience. The CIRCLE program considers this critical in using research evidence to inform decision/policy making. The program's support for The Hydropower Energy Conference held in Sunyani from 12 to 14 Nov. 2015 is a demonstration of commitment toward research uptake and utilization of science for decision making.

Whereas all CIRCLE Visiting Fellows are mandated to publish at least one research paper in an internationally recognized peer-reviewed journal, distributing research mainly within the academic community through journals is not considered to be anywhere near exhausting what CIRCLE considers to be effective knowledge dissemination for impact. The program facilitates the type of research communication that makes known research outputs to a wide range of stakeholders and getting feedback from them. The Research Uptake component of the program is to help the researchers to purposefully design activities to stimulate end users to awareness of, access to, and application of scientific knowledge

The conference focusing on "Hydropower Energy for Economic Growth and Development in West Africa" came at a time that the host country, Ghana, was going through a serious electricity crisis.

At the conference, several initiatives were introduced such as the application of Computational Fluid Dynamics (CFD) for urban water management and Small Hydropower development for economic growth and development in West Africa. The experts and practitioners came together to deliberate on key challenges of Hydropower development and management and reviewing existing prospects of Hydropower generation in West Africa.

7. CONCLUSION

The CIRCLE program prioritizes studies that produce solutions for the immediate needs of countries. Producing knowledge to address some of the challenges brought about by Climate Change and to take advantage of opportunities climate change brings has been addressed by the CIRCLE program at different levels. At the individual level, the program has developed the human capacity to take on the challenges of climate change. Individual researchers have gained more understanding of the concepts in climate change and how they are applied to their respective fields of study. Institutions have been assisted to develop their research systems and have a cadre of ECRs committed to researching climate change issues in agriculture and food security, water, energy, health and livelihoods, and political economy.

REFERENCES

Andrews, K., 2012. Knowledge for Purpose: Managing Research for Uptake—A Guide to a Knowledge and Adoption Program. Department of Sustainability, Environment, Water, Population and Communities, Canberra. https://www.environment.gov.au/system/files/resources/7fee4ddb-e1df-4d13-85f0-e0091a95d80f/files/knowledge-purpose.pdf.

AOSTI (African Observatory of Science, Technology and Innovation), 2014. Assessment of scientific production in the African Union, 2005–2010. http://aosti.org/index.php/report/finish/5-report/15-assessment-of-scientific-production-in-the-african-union-2005-2010.

Garavan, T., McGuire, D., 2001. Competencies and workplace learning: some reflections on the rhetoric and the reality. Journal of Workplace Learning 13 (4), 144–164.

Seleim, A., Ashour, A., Bontis, N., 2007. Human capital and organizational performance: a study of Egyptian software companies. Management Decision. 45 (4), 789–801.

Quitzow, R., Roehrkasten, S., Jacobs, D., Bayer, B., Jamea, E.M., Waweru, Y., Matschoss, P., 2016. The Future of Africa's Energy Supply–Potentials and Development Options for Renewable Energy. IASS Study, Institute for Advanced Sustainability Studies (IASS), Potsdam. http://www.iass-potsdam.de/sites/default/files/files/study_march_2016_the_future_of_africas_energy_supply.pdf.

Chapter 3

Hydropower and the Era of Climate Change and Carbon Financing: The Case From Sub-Saharan Africa

Amos T. Kabo-bah, Caleb Mensah
University of Energy and Natural Resources, Sunyani, Ghana

Chapter Outline

1. Introduction	37	3. The Case of Ghana	45
2. Hydropower and Climate Change	39	4. Conclusions	47
2.1 Carbon Financing	42	5. Recommendations	48
2.2 Hydropower as a Clean		References	49
Development Mechanism Tool	43	Further reading	51
2.3 Carbon Markets and			
Hydropower	44		

1. INTRODUCTION

Increase in the concentration of greenhouse gases and aerosols (microscopic airborne particles or droplets) and variations in solar activity can alter the earth's radiation budget and hence climate. Carbon dioxide (CO_2) has risen by about 30% and is still increasing at a rate of 0.4% per year. This increase is anthropogenic because the changing isotope composition of the atmospheric CO_2 betrays the fossil origin of the increase. These changes in greenhouse gases and aerosols, taken together, are projected to lead to regional and global changes in climate and climate-related parameters. Global models project an increase in global mean temperature of about 1–3.5°C by 2100 and an associated increase in sea level rise of about 15–95 cm. It is projected in Africa that about 75–250 million people will be exposed to increased water stress (IPCC, 2007). The use of coal, gas, and light crude oil (LCO) in power production has been a major source of greenhouse gases, which is increasing the rate of global warming and climate change. It has therefore been suggested that the world turn to cleaner

Sustainable Hydropower in West Africa. https://doi.org/10.1016/B978-0-12-813016-2.00003-4

sources of power production such as wind, solar, geothermal, and nuclear energy. Climate change is a global problem and Africa currently contributes only a small fraction of the emissions that cause climate change. Nevertheless, Africa has an important role for mitigating climate change and thus reducing the adverse effects of climate change on the achievement of the SDGs.

As such, there is a growing need for clean energy technologies throughout the world due to a global decline in fossil-fuel reserves, which results in greenhouse gas emissions leading to global warming (Asumadu-Sarkodie and Owusu, 2016). Sub-Saharan Africa accounts for 13% of the world's population, but only 4% of its population have access to energy. In June 2015, the G7 declared its intention to phase out the use of fossil fuels by the end of the century, affirming the need for radical decarbonization in order to respond to climate change. International climate change negotiations continue to be at the heart of this global response. Consequently, at the heart of these climate negotiations has been fundamental issues of securing national level action, developing carbon markets and other sources of finance, and facilitating the development and transfer of technology.

Newell and Bulkeley (2017) reveal how recent research on energy system transformations in sub-Saharan Africa (SSA), particularly in Kenya, South Africa, and Mozambique, is experiencing competing energy trajectories supported by different actors, along more or less carbon-intensive lines. Moreover, across Kenya, Mozambique, and South Africa, there are different levels of connection to the market, finance and technology mechanisms, which are associated with the climate regime, allowing for further comparison of the extent to which the regime can support "local" decarbonization strategies across diverse institutional settings and uneven levels of economic development (Dubash et al., 2013; Pegels, 2014). As explained earlier, under the international climate regime, the route toward mitigation for countries beyond the Organization for Economic Co-operation and Development (OECD) is further seen to lie in these three main thematic areas: the development of national policy responses, the emergence of carbon markets and other forms of climate finance, and mechanisms for technology transfer (UNFCCC, 2014; Technology Executive Committee, 2011).

Establishing effective carbon markets and forms of finance has been a long-standing tenet of the international policy regime. Negotiations in Paris at COP21 placed a large emphasis on new market mechanisms. The Green Climate Fund has been central to such discussions, resulting from the Copenhagen Accord whereby developed countries committed to jointly mobilizing US$ 100 billion per year by 2020 to address the needs of developing countries—including both public and private funds for climate change mitigation and adaptation. This is alongside the launch of new regionally relevant initiatives in Paris for financing low-carbon energy such as the Africa Renewable Energy Initiative (AREI), which aims to build at least 100 GW of new and additional renewable energy generation capacity by 2020 and 300 GW by 2030. Technology transfer has similarly been central to the development of the international climate policy regime over the past two decades and since Conference of the Parties 13 (COP 13) in Bali, technology transfer and the principle that less economically developed countries should benefit from low-cost, clean technologies as they seek to reduce their emissions has also become one of the pillars of the climate change regime (OECD, 2013; World Bank, 2012). The Clean Development Mechanism (CDM) has helped finance more than 2000 hydropower projects, representing the largest source of OECD bilateral funding for hydropower. (Winkler and Dubash, 2015).

2. HYDROPOWER AND CLIMATE CHANGE

Currently, concerns about environment and climate change management influence choices investors and international financing institutions make concerning energy projects (Bauen, 2006). Since energy is sourced and processed into a usable form from the environment, activities pertaining to its extraction, transportation, conversion, and utilization affect the environment. These impacts are pronounced in thermal energy systems. In a typical fossil-fuel energy system, it is not possible to totally avoid emissions and environmental setbacks because of combustion. Some of the gaseous products of this combustion process are harmful to life and climate system. Therefore, it is important to ensure that energy is extracted, converted, and utilized sustainably. Thus, the sustainability of the environment counts a lot when it comes current energy generation and how the climate system is not compromised in any negative way. The other drivers for sustainable energy include the call for increased access to modern forms of energy especially in the least developed countries to foster development and curb natural resource degradation. Developing countries such as those found in sub-Saharan African (SSA) region are well positioned for the application of renewable energy systems because of the relatively huge demand of sustainable energy for development. Renewable energy supply in the form of electricity and heat may help in alleviating the problems of serious electricity shortage and region's overdependence on traditional biomass (Kaunda et al., 2012a).

Hydropower is stated to be the largest renewable energy resource in the world. According to studies conducted by both the International Energy Agency (IEA) in Florini (2011) and the International Renewable Energy Agency (IRENA) (2012), hydropower produced 3329 TWh of electricity representing a share around 16.5% of the world's electricity in 2009 alone. It is now seen as one of the most important sources of power in many countries. Thus, according to World Energy Council (2010) report, about 160 countries in the world have hydropower in their national electricity generation mixes.

Hydraulic energy in the water is derived from a hydrological cycle. In the hydrological cycle, water constantly flows through a cycle in different phases, evaporating from lakes and oceans, forming clouds, precipitating as rain or snow, then flowing back down to the ocean, seas, dams, rivers, and other water bodies. The main source of energy driving the hydrological cycle is solar, and it is estimated that about 50% of all solar radiation reaching the earth is used to evaporate water in the cycle (IPCC, 2012). Since the hydrological cycle is an endless process, hydropower is considered as a renewable energy resource. The world hydropower potential is very large considering availability of numerous small hydropower potential sites in many countries and potential of water current in rivers and canals (such as water supply and irrigation canals). For a hydropower system, capacity factor depends not only on the availability of water for power generation, but also on whether the power station is designed as a peaking or a base load plant. A peaking plant has low-capacity factor because the plant operates only during specific times while a base-load has a high-capacity factor because the plant operates most of the times. In Africa, despite having small levels of installed capacity, most countries in the region have hydropower in their electricity generation mix. In 2008, hydropower accounted for about 70% of the total electricity generated in the sub-Saharan African region, excluding South Africa (Eberhard et al., 2010). In 2010, 32% of the African's electricity generation capacity was supplied from hydropower (IHA, 2011) (Fig. 1).

Per studies conducted by Kaunda et al. (2012a,b), Howarth et al. (2011), Fearnside (2004), Louis et al. (2000), Kemenes et al. (2007), and Rosa et al. (2004), greenhouse gas (GHGs) emissions from a large-scale storage hydropower station are relatively low as compared to other electricity technologies. The life-cycle GHG emission factors for hydropower technologies are around 15–25 g CO_2 equivalent per kWhel. These are very much less than those of fossil-fuel power generation technologies, which typically range between 600 and 1200 g CO_2 equivalent per kWhel (Lenzen, 2008). Reservoirs in tropical environments have been found to have significant amounts of GHG emission levels than those located in temperate climatic zones. One possible reason for this is the relatively high values of water temperature in tropical climates, which increase the rate of anaerobic organic matter decomposition in the reservoirs (Howarth et al., 2011). Even if the maximum value of GHG emission is considered, emissions from hydropower are very much less than those from fossil fuels. This shows the importance of hydropower in mitigating climate change.

The environment can also impart negative consequences on hydropower generation. For a hydropower project, the quantity of fresh water to generate power is sensitive to the environment and weather in the catchment area. In many cases, environmental degradation in the catchment area is greatly as a result of unsustainable agricultural practices and the use of inorganic fertilizer and unsustainable harvesting of forests. Sedimentation and aquatic weed infestation of reservoirs and rivers may also cause major problems with hydropower generation as a result of environmental degradation. Hydropower resource potential is sensitive to climate change because of its dependence on run-off water, a resource that is dependent on climate-driven hydrology. With global warming, the levels and duration of precipitation are affected. Run-off depends on meteorological parameters such as precipitation and temperature. Thus, the increase

FIG. 1 Map of Ghana showing the transmission network. *(Courtesy: Volta River Authority)*

in global temperature has an effect in water loss through evaporation as well as snow and glacier melting. Studies using the global circulation reveal that, in future, some parts of the world will experience increased run-off, while others reduced run-off as a result of global warming. It is projected that much of Southern and West Africa will have a reduction while East Africa is projected to have increased generation potential (Hamududu and Killingtveit, 2012). From these studies, it can be concluded that if individual countries and regions will experience significant changes in run-off, climate change may not lead to significant changes in the global hydropower generation potential. As most of these studies used global circulation model, they are less reliable despite being able to downscale the modeling to national levels (Kaunda et al., 2012a,b). Notwithstanding that fact, there are likely to be impacts on the system operation, which may require adaptation measures if the existing hydropower systems are to cope with the climatic changes, considering the long life span of most large-scale hydropower projects. Already, extreme weather events like droughts, floods, and hailstorms impact negatively on the hydropower generation by affecting water quantity and quality as well as destroying hydropower plant infrastructure. The other adaptation measures include employing turbine technologies that can optimally be operated in a variable flow environment, and also in a poor water quality environment. Climate change technical adaptation measures, such as flood attenuation and sediment extraction designs, must be integrated in the design of hydropower hydraulic structures like dams, barrages, weirs, settling basins, and channels (Martin et al., 2010).

2.1 Carbon Financing

There are various international funds aimed at climate change, some of them also funding Renewable Energy. The Green Climate Fund (GCF) is planned to become the most important one. It was established by the Cancun Agreement in 2010 in order to finance climate change mitigation activities and is not yet operational. Before the establishment of this financial vehicle, a total of 15 climate funds were already available for Africa (Afful-Koomson, 2015). The World Bank is lending for climate change mitigation directly and manages funds for this purpose. These funds are the Climate Investment Funds (CIF) and the Strategic Climate Fund (SCF) with two subfunds each as well as the most recent Carbon Partnership Facility (CPF). It was the first institution that introduced Green Bonds to fund climate change mitigation. In 2014, the World Bank financing for climate investment increased by $11.3 billion. This amount represents 220 climate projects in 60 countries. The African Development Bank (AfDB) offers a whole range of options for African investors seeking to finance RE projects. The most direct form is the annual energy portfolio, which reached $2 billion for funding clean energy investments on the continent (Schwerhoff and Sy, 2016).

The Climate Investment Funds (CIF), which was designed to help developing countries mitigate greenhouse gas emissions and fight climate change, is funded by the World Bank and the regional development banks, including the AfDB. This fund comprises two climate funds: the Clean Technology Fund (CTF) and the Strategic Climate Fund (SCF). About 42% of CTF funding is spent in Africa. The CIF is an example of funds financed by multilateral donors, which form the Global Environment Facility Trust Fund (GEFTF) and the Global Energy Efficiency and Renewable Energy Fund (GEEREF). This aims to help emerging and developing countries to meet the goals sets by the United Nations Framework Convention on Climate Change (UNFCCC) to mitigate and to adapt to climate change (UNEP, 2012).

2.2 Hydropower as a Clean Development Mechanism Tool

Clean energy sources are urgently needed. Carbon finance has helped to support hydropower, which remains as one of the most scalable short-medium term clean energy technologies available to many countries. Hydropower offers a proven low-carbon technology that can help transform global energy systems, supporting energy access for all while mitigating dangerous climate change. However, badly planned or sited dams can harm local communities and ecosystems, and have triggered longstanding battles for social justice.

Hydropower is the rate at which hydraulic energy is extracted from a specific amount of falling water as a result of its velocity or position or both. The rate of change of angular momentum of falling water or its pressure or both on the turbine blade surfaces creates a differential force on the turbine runner thereby causing rotary motion. As a working fluid, water in a hydropower system is not consumed; it is thus available for other uses. Hydropower can be used to power machinery or to generate electricity or both at the same time. The mechanical application is mainly true for small-scale hydropower plants where the power generated is used to power small-scale mechanical tools and machines for pressing, milling, grinding, and sawing applications. In some instances, the output shaft from the small-scale hydropower turbine is extended in both directions to provide space for both mechanical power provision and electricity generation. Large-scale hydropower plants are normally used for electricity generation (Kaunda et al., 2012a,b).

The Clean Development Mechanism (CDM) as explained by Soanes et al. (2016) is one of the flexible mechanisms of the Kyoto Protocol, started with the setup of the Executive Board in November 2001 and registered the first CDM project in November 2004. The CDM focuses on two objectives, which is the reduction of greenhouse gases (GHG) and contribution to the sustainable development (SD) of the host country. Sustainable development in the CDM means that projects have to be implemented in a sustainable manner as to avoid negative environmental, social, and economic impacts (The Gold Standard, 2003).

Hydropower projects have mostly a bigger impact on social, environmental, and economic issues than other CDM projects. Worldwide hydropower contributes 19% of the total primary energy production (IHA, 2003). 95% of this energy is produced by facilities larger than 10 MW, leaving 5% for small-scale hydropower facilities (Wuebbles et al., 2001). Hydropower is economic in regions with high annual precipitation and cool climates, which reduce evaporative losses and make it possible to store water naturally. Large hydropower projects have received the highest confirmed carbon offsets in the CDM pipeline, producing 48% of renewable energy Certified Emission Reductions (CERs). If emission reductions continue as the CDM pipeline implies, hydropower will overtake industrial gas projects as the largest mitigation tool under the CDM by 2020. Notably, investment trends have remained similar to foreign direct investment, with 78% of confirmed hydropower CERs produced in China, while the least developed countries (LDCs) have received only 1% of this carbon finance (IRN, 2004).

Transforming to low-carbon energy will require substantial public and private finance to be mobilized, especially into developing countries where low-carbon capital-intensive energy technologies often compete poorly with cheaper and presently less risky fossil fuels. Over the past decade, carbon finance has evolved into a new source of funding, representing over 13% of renewable energy investment (Schmitz, 2006). This has been spearheaded by the Clean Development Mechanism (CDM), which lets industrialized countries invest in sustainable low-carbon projects in the developing world as a way to offset their own emissions. Currently, Eighty percent of hydropower potential is yet to be exploited, mostly within emerging and less developed economies, and still represents one of the most scalable short-medium term renewable energy technologies for many countries. Hydropower also offers numerous additional benefits through the integration of other intermittent renewable technologies (Schmitz, 2006).

Despite these potential benefits, there are concerns over the environmental and social (E&S) impacts of "large" hydropower schemes. Failure to manage these impacts is often the primary reason for opposition from local communities and environmental organizations. This is particularly the case with hydropower within developing countries. To be fully sustainable, all hydropower projects require inclusive and holistic engagement with best E&S processes from the start. The international public sector can play a key role in helping developing countries foster responsible private investment.

2.3   Carbon Markets and Hydropower

Carbon markets have already been subject to a host of policy and regulatory changes, making them extremely complex systems (UNFCCC, 2014; Technology Executive Committee, 2011). These have been composed of two main approaches: allowance and project-based systems. Allowance-based systems work where greenhouse gas (GHG) emissions are regulated under a cap that determines how many carbon allowances each entity, region, or country is allowed to emit. Those governed by the scheme are generally allowed to trade their allowances, to let them meet the cap in the most efficient way possible. Project-based systems have, to date, acted through the Kyoto Protocol Flexibility Mechanisms and the voluntary carbon market, whereby GHG emission abatement or sequestration projects are developed, receiving carbon offsets (credits) for what would have occurred under "business as usual." The two most prominent project-based mechanisms are the CDM, which generates Certified Emission Reduction credits (CERs) within developing countries (non-Annex countries of the Kyoto Protocol); and Joint Implementation, which generates Emission Reductions Units (Ockwell et al., 2008). Climate change and carbon market finance under the UNFCCC can help to make lower-carbon pathways more attractive by reducing the risks for investors, offering lines of otherwise unavailable credit and funding projects that are additional (Newell and Bulkeley, 2017).

3. THE CASE OF GHANA

Ghana is a West African country with an agrarian economy and an estimated population of about 25 million. With a lower middle-income status, the country has a nominal GDP of $50 billion and GDP per capita of $1902 (Ghana Statistical Service, 2011). Average annual GDP growth has topped 8% for the past 6 years reaching as high as 14% in 2014. With such impressive economic figures, it is to be expected that demand for energy to power this growth will be high.

Supply of electricity has been unable to meet the demand due to high dependency on rain-fed hydropower plants, which started operating in 1965 and currently account for about 68% of the total installed capacity. The government in complementing the hydropower systems in 1997 installed thermal plants based on light crude oil. However, high crude oil prices on the international market in recent times have made the operation of these thermal plants very expensive, with attempts made to install combined cycle power-generating plants that use liquefied petroleum gas (LPG), but this has also been hindered by problems with the supply of gas from Nigeria.

Currently, Ghana has a total electricity-generating capacity of 3200 MW made up of hydro, thermal power plants, and other renewable energy resources such as solar PV and modern biomass that can be exploited for electricity production. As of 2015, the country's electricity demand was between 14,000 and 16,400 GWh while its available supply is approximately 15,000 GWh. Total hydro is 1580 MW representing about 49.9% as compared to other sources of electricity in the country. Hence, total installed capacity in Ghana is approximately 3167 MW as clearly shown in Table 1.

According to the World Economic Outlook of the International Monetary Fund, GDP growth will be above 5.5% per year from 2015 to 2019 in sub-Saharan Africa. The shared socioeconomic pathways (SSPs) expect in a "middle-of-the-road scenario" that GDP in sub-Saharan Africa will grow with an average annual rate of 3.5% until 2100, so that it almost reaches the development level of the United States today (Leimbach et al., 2015). Given this evidence, Africa has a high level of control over how much climate impacts it will have to face.

In sub-Saharan Africa, a multiplicity of energy systems concurrently provide energy services, including those based on fuel wood, kerosene, small-scale

TABLE 1 Ghana's Total Installed Electricity Capacity as at 2015 (Volta River Authority, 2015)

Plants Installed	Capacity (MW)	Type	Fuel Type
Akosombo	1020	Hydro	Water
Kpong	160	Hydro	Water
Bui HEP	400	Hydro	Water
TAPCO (T1)	330	Thermal	LCO/gas
TICO (T2)	220	Thermal	LCO/gas
Osagyefo power barge	125	Thermal	Gas
T3	132	Thermal	LCO/gas
TT1PP	126	Thermal	LCO/gas
TT2 PP	50	Thermal	DFO/gas
CENIT Energy Ltd	126	Thermal	LCO/gas
MRP	40	Thermal	DFO
Sunon Asogli	200	Thermal	Gas
Karpower barge	225	Thermal	Gas
Genser power	5	Thermal	Gas
VRA solar plant	2.5	Renewable	Solar
Noguchi solar	0.72	Renewable	Solar
Other solar (offgrid and net metered)	3.8	Renewable	Solar
Juabeng oil mill biomass	1.2	Renewable	Bioenergy
Total installed capacity	3167		

renewables, large-scale hydro, and coal-fired generation (particularly in South Africa), while there is significant investment in the development and export of fossil-fuel resources including gas and oil. Despite this, access to energy services is the lowest in sub-Saharan Africa as of any world region, with significant implications for reaching the new sustainable development goal to "Ensure access to affordable, reliable, sustainable and modern energy for all."

Furthermore, with the population expected to both grow and urbanize over the coming century, the International Energy Agency predicts that energy demand will grow by around 80% by 2040 in SSA (IEA, 2014), much of which will be met by the expanded use of fossil fuels unless incentives are put in place to pursue alternative pathways. There has been growing interest in how to support the development of low-carbon energy transitions where measures "to address energy poverty can also be those that would set countries on the much-sought alternative path to low-carbon development" (Christian Aid, 2011). There is also a strong donor discourse about "climate-compatible" development, which advocates interventions that deliver the triple win of poverty alleviation, climate adaptation, and climate mitigation (Mitchell and Maxwell, 2010), mirrored in the concept of transformative change for low-carbon transitions (Winkler and Dubash, 2015).

4. CONCLUSIONS

Sustainable development is development toward a better social, economic, and ecologic situation in a country. Sustainability requirements are often seen as constraints for project developers, arguing that hydropower development will not be financially attractive anymore. The global energy sector over-relies on fossil fuel and is responsible for most of global environmental degradation and climate change. The global demand for energy supply is increasing and the supply is not sustainable. Therefore, fuel substitution to clean energy sources, such as renewable energy, is required. Fuel substitution is the major way to mitigate problems associated with fossil-fuel supply. Hydropower, as reviewed from most literature in this work, is the most feasible source of renewable energy to provide significant levels of global energy, especially electricity. The fact that most of undeveloped potential is located in regions where electricity is needed most, such as in Africa, makes development of hydropower for sustainable energy supply in those regions relevant. Hydropower is one of the most efficient power generation technologies. It is used in many countries and some developed countries solely rely on hydropower for power supply. Therefore, the technology is mature and reliable as well.

Considering the financial constraints in many developing countries for large-scale hydropower projects, small-scale projects may be one of the solutions to the small development of hydropower in such countries. Further, small-scale hydropower technology has the advantage of being applied as a standalone energy system for rural power supply. Therefore, hydropower can significantly contribute toward increased national energy access and security, mitigation of climate change and reduction of harmful air pollutants, creation of economic opportunities, and, thus, effectively leading to sustainable development.

Another general view is that ecologic and economic goals are contrary to one another. A classic example of this conflict is the amount of rest water in diversion-type plants. The more water remaining in the natural river bed and assuring a good ecological condition, the less can be diverted by the channel and goes through the turbine, creating power. Some projects that would have been technically and economic feasible in the past have turned out to be infeasible once delays and compensation costs have been factored in IHA (2003). Costly measures of sustainable planning include good management of the environment, which in the long run will benefit the community, the project, and the nation as a whole (IHA, 2003). Hydropower can make use of local equipment and human skills in building and maintaining a project, in general about 80% local resources in developing countries (IHA, 2003). Run of the river projects are usually cheaper and ecologically sound than projects with dams (Höffken, 2016).

Performing sustainable hydropower could fail to consider one question that every country has to answer first before thinking about new hydro projects as suggested by Schwerhoff and Sy (2016). This relates to the fact that even a very "green" project has an impact on nature and has led the expert group of the LIHI (2004) certification to state, "considerable concerns that 'green' power certification might result in accelerated development of undisturbed river systems." Ecologic sustainability of river hydropower is defined by Bratrich and Truffer (2001) as sustainable hydro, which secures that the river's principal ecologic functions are preserved.

5. RECOMMENDATIONS

A sustainable hydropower project is possible, but needs proper planning and careful system design to manage the challenges. Well-planned hydropower projects can contribute to supply sustainable energy. Further, investments in renewable energy technologies can help create jobs and attract extra income from international carbon trading schemes such as the CDM. Climate change is real and total environmental degradation in the catchment area is unavoidable. Though, on global basis, hydropower potential is projected to increase slightly with global warming, on country level, the situation is projected to be different from one country to the other. Some countries will experience increases in potential while others decrease, but with a great degree of risks in both cases. Therefore, hydropower designs should incorporate adaptation measures. This is an area that should be exploited by further research. Some measures of adaptation concerning variable flow turbine design, incremental power generation, and flood attenuation designs have been stated in the chapter. Synergies concerning clean power production and climate change response offered by hydropower project must also be exploited. For example, apart from hydropower generation, reservoirs can also be used to control floods (one of the climate change adaptation measures).

An appropriate understanding and awareness of the complex technical, environmental, and social issues inherent in a hydropower project requires far more comprehensive environmental and social studies, and this in turn has increased both project costs and lead times. Following the reports from the International Hydropower Association, sustainable planning will finally lead to financial benefits, due to better project acceptance and less delay of the project. As such, it's highly recommended that for large hydropower projects or projects in sensitive areas, full Environmental Impact Assessment is required. African governments should improve their ability to finance crucial projects for the future of their populations.

REFERENCES

Afful-Koomson, T., 2015. The green climate fund in Africa: what should be different? Clim. Dev. 7 (4), 367–379.

Asumadu-Sarkodie, S., Owusu, P.A., 2016. A review of renewable energy sources, sustainability issues and climate change mitigation. J. Cogent Eng. 3 (1), https://doi.org/10.1080/23311916. 2016.1167990.

Bauen, A., 2006. Future energy sources and systems—acting on climate change and energy security. J. Power Sources 157 (2), 893–901.

Christian Aid, 2011. Low Carbon Africa: Leap-Frogging to a Green Future. Author, London.

Dubash, N.K., Raghunandan, D., Sant, G., Sreenivas, A., 2013. Indian climate change policy: exploring a co-benefits based approach. Econ. Polit. Wkly. 48 (22), 47–61.

EAWAG, Bratrich, C., Truffer, B., 2001. "Green" electricity certification for hydropower plans, concept, procedure, criteria, Kastanienbaum. June.

Eberhard, A., Foster, V., Briceno-Garmendia, C., Ouedraogo, F., Camos, D., Shkaratan, M., 2010. Underpowered: the state of the power sector in Sub-Saharan Africa. World Bank Document under the Africa Infrastructure Country Diagnostic Project-Summary of Background Paper 6, http://www.infrastructureafrica.org/system/files/BP6_Power_sector_maintxt.pdf.

Fearnside, P.M., 2004. Greenhouse gas emissions from hydroelectric dams: controversies provide a springboard for rethinking a supposedly "clean" energy source. An editorial comment. Clim. Chang. 66 (1–2), 1–8.

Florini, A., 2011. The International Energy Agency in global energy governance. Global Policy 2 (s1), 40–50.

Ghana Statistical Service, 2011. Ghana Multiple Indicator Cluster Survey with an Enhanced Malaria Module and Biomarker. Final Report Accra, Ghana.

Hamududu, B., Killingtveit, A., 2012. Assessing climate change impacts on global hydropower. J. Energies 5, 305–322.

Höffken, J.I., 2016. Demystification and localization in the adoption of micro-hydro technology: insights from India. Energy Res. Soc. Sci. 22, 172–182.

Howarth, R., Santoro, R., Ingraffea, A., 2011. Methane and the greenhouse-gas footprint of natural gas from shale formations. J. Clim. Change 106 (4), 679–690.

IEA, 2014. Statistics, I. E. A. OECD.

Intergovernmental Panel on Climate Change, 2012. Hydropower. In: Special Report on Renewable Energy Sources and Climate Change Mitigation. Cambridge University Printing Press. New York, Special report of IPCC. (Chapter 5).

Intergovernmental Panel on Climate Change (IPCC), 2007. Climate Change 2007: Impacts, Adaptation and Vulnerability, Contribution of Working Group II to the Fourth Assessment Report of the Intergovernmental Panel on Climate Change. Cambridge University Press, Cambridge.

International Hydropower Association, 2003. The role of hydropower in sustainable development. IHA white paper, February. http://www.hydropower.org/DownLoads/IHA%20White%20 Paper_260203_LowRes.pdf.

International Hydropower Association (IHA), 2011 Activity Report, International Hydropower Association, London, 2011.

International Renewable Energy Agency (IRENA), 2012. Renewable energy technologies-cost analysis series, volume 1: power sector IRENA working paper 3/5.

International River Network, 2004. In: McGully, P., Wong, S. (Eds.), Powering sustainable future: The role of large hydropower in sustainable development. Prepared for the UN Symposium on hydropower and Sustainable Development. IRN.

Kaunda, C., Kimambo, C., Nielsen, T., 2012a. Potential of small-scale hydropower for electricity generation in Sub-Saharan Africa. ISRN Renew. Energy. Article ID 132606, 15 pages.

Kaunda, C.S., Kimambo, C.Z., Nielsen, T.K., 2012b. Hydropower in the context of sustainable energy supply: a review of technologies and challenges. ISRN Renew. Energy. Article ID 730631, 15 pages, https://doi.org/10.5402/2012/730631.

Kemenes, A., Forsberg, B., Melack, J., 2007. Methane release below a tropical hydroelectric dam. Geophys. Res. Lett. 34, Article L12809.

Leimbach, M., Kriegler, E., Roming, N., Schwanitz, J., 2015. Future growth patterns of world regions—a GDP scenario approach. Global Environ. Change 42, 215–225.

Lenzen, M., 2008. Life cycle energy and greenhouse gas emissions of nuclear energy: a review. Energy Convers. Manag. 49 (8), 2178–2199.

Louis, V.L.S., Kelly, C.A., Duchemin, E., Rudd, J.W.M., Rosenberg, D.M., 2000. Reservoir surfaces as sources of greenhouse gases to the atmosphere: a global estimate. Bioscience 50 (9), 766–775.

Low Impact hydropower Institute (LIHI), 2004. Low impact hydropower certification criteria, summary of goals and standards. www.lowimpacthydro.org.

Martin, O., Lillehammer, L., Hveling, O., 2010. In: Hydropower development and curbing climate gas emissions: a win-win opportunity. Proceedings of the 6th International Conference on Hydropower (Hydropower '10), February. International Centre for Hydropower, Tromsø.

Mitchell, T., Maxwell, S., 2010. Defining Climate Compatible Development. CDKN, London.

Newell, P., Bulkeley, H., 2017. Landscape for change? International climate policy and energy transitions: evidence from sub-Saharan Africa. Clim. Pol. 17 (5), 650–663. https://doi.org/10.1080/14693062.2016.1173003.

Ockwell, D., Watson, J., MacKerron, G., Pal, P., Yamin, F., 2008. Key policy considerations for facilitating low-carbon technology transfer to developing countries. Energy Policy 36 (11), 4104–4115.

OECD, 2013. Effective Carbon Prices. Author, Paris.

Pegels, A. (Ed.), 2014. Green Industrial Policy in Emerging Countries. Routledge, London.

Rosa, L.P., dos Santos, M.A., Matvienko, B., dos Santos, E.O., Sikar, E., 2004. Greenhouse gas emissions from hydroelectric reservoirs in tropical regions. Clim. Chang. 66 (1–2), 9–21.

Schmitz, D., 2006. International Climate Policy. Hamburgisches Welt-Wirtschafts-Archive (HWWA) Report, Hamburg Institute of International Economics. ISSN: 0179-2253.

Schwerhoff, G., Sy, M., 2016. Financing renewable energy in Africa—key challenge of the sustainable development goals. Renew. Sust. Energ. Rev. https://doi.org/10.1016/j.rser.2016.11.004.

Soanes, M., Skinner, J., Haas, L., 2016. Sustainable Hydropower and Carbon Finance. IIED, London.

Technology Executive Committee, 2011. Briefing note on the development and transfer of technologies under the UNFCCC process. Retrieved from UNFCCC website: http://unfccc.int/ttclear.

The Gold Standard, 2003. The gold standard clean development mechanism PDD. www.panda.org/goldstandard.

UNEP, 2012. Financing renewable energy in developing countries. United Nations Environment Programme Finance Initiative.

UNFCCC, 2014. 2014 Biennial Assessment and Overview of Climate Finance. Author, Bonn.

Volta River Authority, 2015. Ghana's power outlook—facts and figures. June.

Winkler, H., Dubash, N., 2015. Who determines transformational change in development and climate finance? Clim. Pol. Advance online publication. https://doi.org/10.1080/14693062.2 015.1033674.

World Bank, 2012. Inclusive Green Growth: The Pathway to Sustainable Development. Author, Washington, DC.

World Energy Council, 2010. Survey of Energy Resources: Hydropower. World Energy Council, London, UK.

Wuebbles, D.J., Jain, K.A., Watts, R.G., 2001. Concerns about climate change and global warming. Fuel Process. Technol 71, 99–119.

FURTHER READING

Africa Progress Panel, 2015. Power, People, Planet: Seizing Africa's Energy and Climate Opportunities, Africa Progress Panel, Geneva. http://www.africaprogresspanel.org/wp-content/uploads/2015/06/APP_REPORT_2015_FINAL_low1.pdf.

Ahlborg, H., Sjostedt, M., 2015. Small-scale hydropower in Africa: socio-technical designs for renewable energy in Tanzanian villages. Energy Res. Soc. Sci. 5, 20–33 (Special issue on Renewable Energy in Sub-Saharan Africa Contributions from the Social Sciences).

Fearnside, P., 2002. Greenhouse gas emissions from a hydroelectric reservoir (Brazil's Tucurui dam) and the energy policy implications. J. Water Air Soil Pollut. 133 (1–4), 69–96.

Inter-Academy Council, 2007. Energy supply. In: Lighting the Way: Toward a Sustainable Energy Future Report. The Inter-Academy Council, The Netherlands (Chapter 3).

International Energy Agency, 2012. 2011 key world energy statistics. IEA report, The Economic Co-Operation and Development (OECD).

UNFCCC, 2012. UNFCCC releases report on the benefits of the Kyoto Protocol's clean development mechanism. Bonn.

UNFCCC, 2015. Adoption of the Paris Agreement, Draft Decision −/CP.21. Author, Bonn.

UNFCCC Press Office, n.d. 20th November. Retrieved February 13, 2014, from http://cdm.unfccc. int/CDMNews/issues/issues/I_KYD6PO19YS9DE7YGH894BJ9WRBCQ4Z/viewnewsitem. html.

Chapter 4

Hydropower Development— Review of the Successes and Failures in the World

Xie Yuebo*, Amos T. Kabo-bah[†], Kamila J. Kabo-bah[†], Martin K. Domfeh[†]
*Hohai University, Nanjing, China
[†]University of Energy and Natural Resources, Sunyani, Ghana

Chapter Outline

1. Introduction	53	4. Lessons From Cited Case Studies	58
2. Methodology of the Study	54	5. Conclusion	60
3. Successful and Failed		References	60
Hydropower Projects	54		

1. INTRODUCTION

A crucial goal among the Sustainable Development Goals is the need to ensure access to modern, affordable, sustainable, and reliable source of energy. Despite the immense successes chalked up over the years in terms of global access to improved energy source, 2.6 billion people are still without any modern form of energy (IEA, 2012). This prevailing situation is particularly alarming due to the direct correlation between energy availability and other key parameters such as economic growth, poverty reduction, education, health, and agriculture.

Sustainable sources of energy mainly encompass renewable energy sources due to their high efficacy in terms of reductions in greenhouse gases, affordability, and supply security (Terrapon-Pfaff et al., 2014). Among the various sources of renewable energy, hydropower alone supplies about 16% of the global energy with expected implementation of more hydropower projects in the coming years due to the immense benefits associated with it (Kumar et al., 2011).

Notwithstanding this fact, sustainable management of hydropower projects continues to be a key challenge to many governments and other stakeholders. The source of this challenge stems from the fact that the development and

Sustainable Hydropower in West Africa. https://doi.org/10.1016/B978-0-12-813016-2.00004-6

management of hydropower projects exhibit a number of complexities: multiple stakeholders with varied interests, hydrologic and climatic uncertainties, huge initial capital, complexity in design and implementation of the project, requires skilled personnel, etc. In the past, most electricity development and management schemes were in the hands of governments. Later experiences and reformations included other stakeholders such as communities and private partners. Experiences from both successful and unsuccessful development and management of these projects are key toward the implementation of future hydropower projects especially for emerging economies especially in the West African Region. This chapter therefore presents the findings of a carefully selected outcome of some selected hydropower projects around the world.

2. METHODOLOGY OF THE STUDY

This study gathers information from sampled literature on hydropower projects in the world. According to Higiro et al. (2015), the indicators for the measure of success or otherwise of a project are often difficult to agree upon; hence, the study employs a holistic view of these projects in terms of the economic, social, and environmental impacts based on what has been reported in literature.

3. SUCCESSFUL AND FAILED HYDROPOWER PROJECTS

According to Pervaz and Rahman (2012), a 70-kW micro hydel project at Trongsa in Bhutan has been operating successfully due to its successful community management and ownership scheme as well as the innovative approach of the project.

A decentralized district energy-planning program in Nepal called the Rural Energy Development Programme (REDP) has implemented a number of community-managed micro hydel plants with funding from the Government of Nepal and United Nations Development Programme (UNDP) with the World Bank as project partner. The focal point of this program is on micro hydel plants even though in some few instances other renewable energy projects such as biogas and solar PV lighting are implemented. The success of this project has been attributed to the following: the focus on socioeconomic development at the district level, utilization of indigenous energy resources, and financial assistance from funding agencies. REDP has been deemed the most successful renewable energy project in Nepal (Acharya, 2008; Pervaz and Rahman, 2012).

A 100-kW micro hydro plant was built in Siklesh village in Nepal through the Annapurna Conservation Area Project (ACAP) at an estimated cost of US$121,755 in 1994. The aim of the project was to provide reliable and modern source of electricity to 346 households thus eliminating the use of kerosene lanterns and enhancing standard of living. This project has been successful due to the inclusion of the concept of community ownership at the onset of the project. This was achieved through a carefully designed implementation policy in which

86% of the funding emanated from donors with the remaining contribution coming from the community through communal labor. Notwithstanding these achievements, the project like many other mini-hydro plant facilities elsewhere has had some challenges. Prominent among them is the limited load factor of the plant that impacts negatively on the commercial viability of the energy facility (Guta et al., 2015; Sovacool et al., 2011; Fulford et al., 2000).

In Pakistan, the Aga Khan Rural Support Programme (AKRSP), which commenced in 1982 as a nonprofit organization, has chalked up a number of successes in terms of the provision of over 180 micro-hydro power plants with capacities ranging from 20 to 75 kW. The focus of the project is poverty reduction in Northern Pakistan. This community-collaborated project has ensured the provision of reliable source of electricity to over 175,000 people (Pervaz and Rahman, 2012).

In Sri Lanka, the Energy Services Development/Renewable Energy for Rural Economic Development (ESD/RERED) has been adjudged as one of the most successful renewable energy projects with coverage exceeding 100,000 households. The project is mainly made up of off-grid micro hydroelectric systems and solar home systems. An extended component of the project also provided 126 MW of grid-connected power plants for 500,000 households. The success of the project stemmed from the minimal utilization of foreign expertise, financial assistance from government via provincial councils, and management of project by private institutions (Pervaz and Rahman, 2012).

The Bujagali Dam situated in Uganda was designed to generate 250 MW of hydroelectric power, but this target could not be actualized due to reduced inflow flow from the Lake Victoria thus rendering the facility inefficient. The reduced inflow has been attributed to fluctuating levels of the Lake Victoria as a result of climate change impacts. The World Bank attributed the problem to lack of assessment of the potential impacts of climate change on the project. Similarly, Nalubaale and Kiira Dams were designed to produce 380 MW of electricity, but the prevailing mean output ranges between 110 and 135 MW (Higiro et al., 2015).

Another relevant experience can be harnessed from the Plan for Renewable Energies and the Rational Use of Energy (PRERURE) and Green Energy Revolution Reunion Island (GERRI) Renewable Energy Projects that were launched in the Island of Reunion in France in 2000 and 2008, respectively. It is the target of these two projects to attain 100% renewable energy source in the energy mix by 2025. This ambition is being aspired through favorable policies such as feed-in-tariffs, tax exemptions, and direct subsidies. The island currently has 146-MW installed hydropower capacity distributed at six locations with an expected implementation of additional 50 MW by 2020 under these two projects (Drouineau et al., 2015; Praene et al., 2012; Lin et al., 2016).

The 400-MW Bui Hydroelectric Power Project is located on the Black Volta Basin of Ghana precisely along the borders of the Brong Ahafo and the Northern regions. The estimated cost of the project was US$622 million: US$60 million came from the Government of Ghana, US$263.5 million from a concessional loan, and US$298.5 million was provided by the Chinese Exim-Bank. The projected was commenced in 2009 and successfully commissioned in 2013. The creation of the dam resulted in the relocation of 1216 people and inundation of about 21% of the Bui National Park. The inundation of the park threatened the habitat of over 250 black hippopotamus; hence, adequate measures were put in place to relocate these animals. There are plans to include a 30,000-ha irrigation project and a 250-MW Solar Farm Project has commenced. The hydropower project has contributed to the socioeconomic development of the indigenes through the provision of schools, roads, employment, social amenities, and resettlement of the 1216 displaced people (Bui Power Authority, 2017; Government of Ghana, 2017; Graphic Online, 2017).

The Akosombo Dam whose construction was completed in 1965 has for the past years supported the power industry in Ghana. The facility was constructed at an estimated cost of £130 million with the primary aim of providing electricity to the aluminum industry in addition to other purposes. The funding agencies included the World Bank as well as the United States and the British governments. The facility also supplies electricity to other neighboring countries (Togo, Benin, and Côte D'Ivoire) through a subregional interconnected grid. Despite the untold socioeconomic benefits of the Akosombo Hydropower Project, the facility has also brought in its trail a number of negative consequences: displacement of 80,000 indigenous inhabitants, widespread outbreak of malaria, river blindness outbreak and bilharzia, reduced floodplain agriculture, reduced fishing activities, growth of aquatic weeds, formation of sandbar at the estuary as well as promotion of coastal erosion (Kalitsi, 2003; Miescher, 2014; Mul et al., 2015).

The proposed construction of the Tipaimukh Dam in Manipur, India, was halted due to a number of intense protest from civil groups, NGOs, fringing communities, and researchers who cited a number of reasons. The reasons cited by the protesters included inundation of about $275\,km^2$ of forest land, displacement of about 60,000 local people, threats on food security, and the vulnerability of the region to earthquakes (Greyling et al., 2014; Paudyal and Panthi, 2010).

Similarly, the proposed construction of a Tiger Leaping Gorge Dam in Yunnan in China was stopped in the midst of severe protest by civil groups, NGOs, and key scientific experts. The success of their protest arose from the fact that not only did they elaborate on the cultural heritage and environmental concerns of the proposed project, but also they provided alternative places of locating the dam that will result in less impact (Greyling et al., 2014; International Rivers, 2012).

However, in relation to some other hydropower projects, the use of anti-dam protests from civil groups, environmentalists, indigenous people as well as nongovernmental organizations (NGOs) failed due to the immense political power behind these projects. Serious concerns related to the absence of an Environmental Impact Assessment (EIA), flawed procurement, and approval procedures caused a temporal halt of the Nam Mang 3 Hydropower Project in Laos, China. These concerns brought in the intervention of the World Bank,

Asian Development Bank, and the International Monetary Fund (IMF). But these challenges could not stop the project from being implemented due to the unflinching quest and determination on the part of the Chinese Government to promote economic development. This resulted in the successful completion of the project in 2005 (Greyling et al., 2014; Hydropower Kunming Engineering Corporation, 2012; International Rivers Network, 2013; Phouthonesy, 2012).

Similarly, the Pubugou Hydropower Project situated in Sichuan China was successfully completed despite the controversial saga that bedevilled the project. There was a massive protest by the indigenes due to poor involvement of the local people in decision making pertaining to the project. The protest led to the arrest of the main leader of the antidam protest, Chen Tao, in addition to other leaders. Chen Tao was subsequently imprisoned and sentenced to death on the grounds of allegedly killing a police man during the protest. It is obvious that governmental veto and determination were the main factors that ensured the success of the project (BBC, 2006; Beitarie, 2011; Greyling et al., 2014; Mertha, 2008). A summary of the cited cases of hydropower projects in this chapter has been provided in Table 1.

TABLE 1 Cited Cases of Some of the Successful and Unsuccessful Hydropower Projects

Name of Project	Location	Remarks	Factors
Trongsa Micro Hydel Project	Trongsa, Bhutan	Successful	Community management and ownership
Rural Energy Development Programme (REDP)	Nepal	Successful	Focus on socioeconomic development, Utilization of indigenous energy resources, and Financial assistance from funding agencies
Annapurna Conservation Area Project (ACAP)	Siklesh, Nepal	Successful	Community ownership
Aga Khan Rural Support Programme (AKRSP)	Pakistan	Successful	Community collaboration and involvement
Development/ Renewable Energy for Rural Economic Development (ESD/RERED)	Sri Lanka	Successful	Minimal utilization of foreign expertise, Financial assistance from government, and Management of project by private institutions

TABLE 1 Cited Cases of Some of the Successful and Unsuccessful Hydropower Projects—cont'd

Name of Project	Location	Remarks	Factors
Bujagali Dam	Uganda	Unsuccessful	Reduced inflow flow due to climate change
Bui Hydroelectric Power Project	Ghana	Successful	Adequate funding, Community participation, and development and Implementation of appropriate environmental measures
Akosombo Hydropower Project	Ghana	Successful	Adequate funding and the Concept of regional integration and development
PRERURE and GERRI Renewable Energy Projects	Island of Reunion, France	Successful	Feed-in-tariffs, Tax exemptions, and Direct subsidies
Tipaimukh Dam	Manipur, India	Unsuccessful	Environmental concerns, Threats on food security, and Vulnerability of the region to earthquakes
Tiger Leaping Gorge Dam	Yunnan, China	Unsuccessful	Sociocultural and Environmental concerns, Provision of alternative location
Nam Mang 3 Hydropower project	Laos, China	Successful	Governmental veto and determination
Pubugou Hydropower Project	Sichuan, China	Successful	Governmental veto and determination

4. LESSONS FROM CITED CASE STUDIES

A number of lessons could be drawn from the surveyed literature. Firstly, the indispensable significance of energy brings on board various stakeholders and players with varied interests. The success of these projects therefore rests on the ability to meet as much as possible the varied expectations of these various factions. An all-inclusive stakeholder dialogue and involvement throughout the project cycle presents one of the best tools used to meet the varied aspirations of the different stakeholders. It is therefore not surprising that the World Bank now insists on an in-depth social consultation especially with indigenes before implementation of any large hydropower project in order to ensure transparency and sustainability of the project (Higiro et al., 2015).

As already reported by Kalitsi (2003), the unpleasant environmental consequences linked to the development of large hydropower projects as in the case of the Akosombo Hydropower Project could in most cases be mitigated with adequate investigation before, during, and upon completion of the project. Inadequate environmental impact assessment often leads to untold consequences on the environment and the fringe communities.

Among the reviewed papers, it appears obvious that community management of small hydropower plants has generally proven efficient due to the inherent sense of community ownership and participation. When indigenes are made to participate in decision making, implementation, and management of energy projects, they are propelled to offer their very best toward the sustainability of the project due to the inculcated sense of ownership.

Another crucial revelation lies in the relevance of adequate availability of financial aid from funding agents especially for medium- and large-scale hydropower projects. It is an established fact that the World Bank and other funding agencies and governments have been very instrumental in the implementation of countless hydropower projects especially in developing countries. These continuous financial provisions will go a long way to minimize the already-existing gap between developed and developing countries in terms of access to modern and reliable energy.

It is without doubt that every successful project thrives in a conducive and fertile environment. As evident in the PRERURE and GERRI Renewable Energy Projects, such favorable policies for hydropower projects include appropriate tax exemptions, feed-in tariff policies, and implementation of substantial subsidies.

Again, adequate funding from both local and external partners is also very crucial for the success of any hydropower project due to the huge initial capital required. Inadequate availability of funds could result in project delays and undue costs.

Also there is the need to prioritize the possible consequences of climate change and hydrologic uncertainties on energy generation during the planning of hydropower projects. Enhanced research in this area should also be promoted in order to add to the ever-increasing knowledge on the impacts of climatic conditions on hydropower generations. There is also the need for stronger collaboration between engineers and the academia for the development of innovative interventions and approaches to mitigate climate change impacts on energy generation from hydropower facilities.

Finally, the possible impacts of antidam protest by Nongovernmental organizations (NGOs), indigenes, researchers, and civil groups on the success or failure of hydropower projects cannot be overemphasized. Notwithstanding this crucial factor, some governments have often used various kinds of tools and power to nullify these arguments and protests thus paving way for the successful implementation of some of these projects.

5. CONCLUSION

As the world continues to strive toward ensuring improved access to sustainable energy for all the populace, lessons from already-implemented hydropower projects will continue to remain indispensable especially for growing economies. From the surveyed case studies, the authors recommend the following measures for sustainable development of future hydropower projects: stakeholder dialogue and involvement; adequate investigation before, during, and upon completion of the project; community management of small hydropower plants; favorable policies; adequate funding; prioritization of possible impacts of climate change and hydrologic uncertainties on energy generation; and constructive public protest where adverse impacts cannot be mitigated.

REFERENCES

Acharya, M., 2008. Critical Factors in Determining Success of Renewable Energy Projects in Nepal. Report of SAARC Energy Centre.

BBC, 2006. China Executes Dam Protester. http://news.bbc.co.uk/2/hi/asia-pacific/6217148.stm.

Beitarie, R., 2011. Surge of new dams in Southwest China produces power and public ire. In: Circle of Blue. Retrieved from, http://www.circleofblue.org/waternews/2011/world/burst-ofnew-dams-in-southwest-china-produces-power-and-public-ire/.

Bui Power Authority, 2017. http://www.buipower.com/node/143 (Retrieved from 2 August 2017).

Drouineau, M., Assoumou, E., Mazauric, V., Maïzi, N., 2015. Increasing shares of intermittent sources in Reunion Island: impacts on the future reliability of power supply. Renew. Sust. Energ. Rev. 46, 120–128. https://doi.org/10.1016/j.rser.2015.02.024.

Fulford, D.J., Mosley, P., Gill, A., 2000. Field report recommendations on the use of micro-hydro power in rural development. J. Int. Dev. 12 (7), 975–983.

Government of Ghana, 2017. Bui and the Tale of the Three Hydro Dams. http://www.ghana.gov.gh/index.php/media-center/features/805-bui-and-the-tale-of-three-hydro-dams (Retrieved from 2 August 2017).

Graphic Online, 2017. Bui Power Authority Starts 250MW Solar Farm Next Month. http://www.graphic.com.gh/news/general-news/bui-power-authority-starts-250mw-solar-farm-next-month.html (Retrieved from 2 August 2017).

Greyling, T., Knippers, J., Tumakova, Y., 2014. Regional hydropower projects: what can be learned from successful and unsuccessful public resistance? In: Senz, A., Reinhardt, D. (Eds.), Task Force: Connecting India, China and Southeast Asia—New Socio-Economic Developments, pp. 24–33. No. 97/201. Duisburg, Germany.

Guta, D., Jara, J., Adhikari, N., Qiu, C., Gaur, V., 2015. ZEF-discussion papers on development policy no. 203 decentralized energy in water-energy-food security nexus in developing countries: case studies on successes and failures. p. 203.

Higiro, G., Mbabazi, P., Kibachia, J., 2015. Influence of implementation factors on effective delivery of energy projects in Rwanda: case of Nyabarongo I hydro electric power project. Int. J. Bus. Manag. 3 (9), 251–260.

Hydropower Kunming Engineering Corporation, 2012. Nam Mang 3 Hydropower Project in Laos. Retrieved from, http://www.khidi.com:8083/KHIDI2009/KD2011E/K_Article.asp?ListName=Overseas0AProjects&ids=55145&DataBaseName=XxzxMainMsg.

International Energy Agency (IEA), 2012. World Energy Outlook 2012: WEO2012 Measuring Progress towards Energy for All. IEA, Paris.

International Rivers, 2012. Jinsha River Dams, Jinsha River (Upper Yangtze River) Hydropower Projects List. Retrieved from, http://www.internationalrivers.org/resources/jinsha-riverdams-3604.

International Rivers Network, 2013. Report: New Lao Dam Embroiled in Controversy.

Kalitsi, E.A.K., 2003. In: Problems and Prospects for Hydropower Development in Africa. The Workshop for African Energy Experts on Operationalizing the NGPAD Energy Initiative. Novotel, Dakar, Senegal. pp. 1–25.

Kumar, A., Schei, T., Ahenkorah, A., Caceres Rodriguez, R., Devernay, J.-M., Freitas, M., Liu, Z., 2011. Hydropower. In: IPCC Special Report on Renewable Energy Sources and Climate Change Mitigation, pp. 437–496 (Chapter 5).

Lin, J., Wu, Y., Lin, H., 2016. Successful experience of renewable energy development in several offshore islands. Energy Procedia 100 (September), 8–13. https://doi.org/10.1016/j.egypro.2016.10.137.

Mertha, A., 2008. China's Water Warriors: Citizen Action and Policy Change. Cornell University Press, Ithaca, NY.

Miescher, S.F., 2014. "'Nkrumah's baby'": The Akosombo dam and the dream of development in Ghana, 1952–1966. Water Hist. 6, 341–366. https://doi.org/10.1007/s12685-014-0112-8.

Mul, M., Sidibé, Y., Annor, F., Ofosu, E., Boateng-Gyimah, M., Ampomah, B., Addo, C., 2015. Balancing hydro generation with sustainable ecosystem management. In: Water Storage and Hydro Development for Africa. Palais des Congrès de la Palmeraie, Marrakesh, pp. 46–50.

Paudyal, H., Panthi, A., 2010. Seismic vulnerability in the Himalayan region. In: Himalayan Physics pp. 14–17.

Pervaz, M., Rahman, L., 2012. In: Review and evaluation of successful and unsuccessful renewable energy projects in South Asia. 2012 International Conference on Life Science and Engineering, 45. pp. 6–11. https://doi.org/10.7763/IPCBEE.

Phouthonesy, E., 2012. Lao PDR: Please Give Us a Chance to Rise above Poverty. In: Energy. Lao People's Democratic Republic. Retrieved from, http://laospdrnews.wordpress.com/%0Acategory/energy/page/2/.

Praene, J.P., David, M., Sinama, F., Morau, D., Marc, O., 2012. Renewable energy: Progressing towards a net zero energy island, the case of Reunion Island. Renew. Sust. Energ. Rev. 16 (1), 426–442. https://doi.org/10.1016/j.rser.2011.08.007.

Sovacool, B.K., Bambawale, M.J., Gippner, O., Dhakal, S., 2011. Energy for Sustainable Development Electrification in the Mountain Kingdom: the implications of the Nepal Power Development Project (NPDP). Energy Sustain. Dev. 15 (3), 254–265. https://doi.org/10.1016/j.esd.2011.06.005.

Terrapon-Pfaff, J., Dienst, C., König, J., Ortiz, W., 2014. A cross-sectional review: impacts and sustainability of small-scale renewable energy projects in developing countries. Renew. Sust. Energ. Rev. 40, 1–10. https://doi.org/10.1016/j.rser.2014.07.161.

Chapter 5

Climate Change and Societal Change—Impact on Hydropower Energy Generation

Mary Antwi, Daniella D. Sedegah
University of Energy and Natural Resources, Sunyani, Ghana

Chapter Outline

1. Introduction   63
2. Climate Change and
   Hydropower Generation of
   Electricity in Sub-Saharan Africa   65
3. Climate Change and Patterns
   of Societal Consumption of
   Generated Hydropower   66
   3.1 Availability of
       Electrical Energy
       Supply in Sub-Saharan Africa 66

4. Climate Change and Societal
   Change in the Face of
   Hydropower Energy Generation   69
5. Conclusion   71
   References   71

1. INTRODUCTION

Resource-rich Africa, according to the African Economic Outlook (OECD, 2016), remained the second fastest-growing economy after East Asia with an estimated growth in real GDP of 3.6%, higher than the 3.1% for global economy. It further estimates that average growth in Africa is expected to remain moderate at 3.7% in 2016 but could accelerate to 4.5% in 2017 depending on the strength of the world economy and a gradual recovery in commodity prices. Africa's increasing economic growth has been driven by high commodity prices and substantial investment from Asian countries, especially China, which is leading foreign investment in African infrastructure. Growth in Sub-Saharan Africa conversely has not been matched by access to modern energy; water and other basic services remain low across the region (IHA, 2016). A challenge to hydropower generation is climate change in terms of changes in rainfall and water availability, protracted drought events, significant variation in historical

Sustainable Hydropower in West Africa. https://doi.org/10.1016/B978-0-12-813016-2.00005-8

temperature regimes, and more frequent and severe weather events, including floods (IHA, 2016). The African Development Bank (2015) intimates that climate change is likely to exert severe pressure on water supplies; this pressure will be exacerbated by Africa's increasing urbanization and industrialization, which has wide-ranging repercussions for agriculture, industry, and sanitation, and in some countries for hydroelectric power generation. Arguments in favor of hydropower as a source of energy are: (1) It is quite reliable and not subject to international price fluctuations that can affect oil and gas markets; (2) Dams for hydropower are also used to store, use, and divert water for consumption, irrigation, cooling, transportation, construction, and recreation; and (3) It is relatively cheaper compared with other electricity sources, particularly so in the case of large hydropower plants (IEA, 2012).

Society adapts to changes as and when it is necessary in order to survive. Climate change has been one of the major influences that cause societies to change their way of living (Shove, 2010). Patterns of consumption of most natural resources are no more seen as sustainable, if new ways of living are not adopted. This is because climate change will continuously influence the way individuals live especially people who depend mostly on the natural resources that are sensitive to climate change (Adger et al., 2003). Weather has a great influence on natural resources, especially water resources that are needed for the production of electricity generated from hydropower. Electrical energy generated from hydropower is considered as renewable, affordable, and environment friendly as compared to other sources of energy generation such as fossil fuels (Berga, 2016). In the developing countries, for example, to ensure that people have access to electricity, there should be affordable and reliable localized ways of generating the power, and that is through hydropower generation (IHA, 2015). The implication of ensuring that energy reaches as many people as possible in developing countries, of which Sub-Saharan Africa is one, is that they should make use of the opportunities that are available to them in terms of water resources to generate electrical energy.

Sub-Saharan Africa is well endowed with water resources (having 50 river basins) (UNFCCC, 2007) that have potentials of generating the needed hydropower (about 10% of the world's total hydropower potential) (Pérez-Sánchez et al., 2017) that could supply electricity for majority of the people. Nevertheless, the region is suffering from low-energy generation and consumption with only about 25% of the population in the region having access to electrical energy (IHA, 2016). The question that arises from this observation of low hydropower generation in a region well endowed with water resources is—why is the continent unable to tap such resources to their advantage? According to a review by Adger et al. (2003), developing countries are at a higher risk to climate change since they are more vulnerable to the effects of climate change. The fact that climate change is happening all over the world suggests that people should consequently cope with the changes and adapt to situations as they are predicted (Adger et al., 2009). The need to adapt arises from the fact that climate change is affecting almost all sectors of the regions across the globe, causing extreme temperatures, droughts, famine, floods, and many other natural hazards, as the population continues to grow (Cobbinah et al., 2015).

Sub-Saharan Africa is faced with some social limits such as economic and technical capacities (Adger et al., 2009) to adapt to this climate change. The region, for the past decades, has been confronted with disease outbreaks, poverty, famine, droughts, and many more setbacks to development. Unfortunately, due to poverty, the region does not have the needed new technologies employed by the developed societies to counter all these vices in order to develop their energy sector through the opportunities that are available to the region (Cobbinah et al., 2015). However, the prospects for developing the energy sector could be revived through public-private partnership. Lessons from successes on hydropower exploitation in the advanced societies could be a starting point.

This review therefore examines the impact of climate change on the water resources for hydroelectric generation in Sub-Saharan Africa, and the response and attention that it is gaining from the society. The capacity of the water resources to generate sustainable electricity to feed the majority of the people in Sub-Saharan Africa is paramount to the changes and development that will occur within the African societies from the energy prospects.

2. CLIMATE CHANGE AND HYDROPOWER GENERATION OF ELECTRICITY IN SUB-SAHARAN AFRICA

Global installed capacity of hydropower development has grown by 27% with an average growth rate of 3% per annum particularly in emerging markets such as African countries (World Energy Resources, 2015). Hydropower has become needful as a result of increase in demand for electricity for purposes of industrialization and manufacturing. In addition to hydropower providing clean energy, it also provides water services, energy security and facilitates regional cooperation and economic development, necessary ingredients for the development of any country. Hydropower is often a major factor in economic and social development at local, national, and regional levels. Water bodies are hardly restricted to one geographical area but cross local and national boundaries, hence involve cooperation not just among domestic stakeholders, but also with stakeholders from neighboring countries as well. A case in point is the River Nile, which is 6853-km (4258 miles) long, with its drainage basin covering 11 countries, namely, Tanzania, Uganda, Rwanda, Burundi, Congo-Kinshasa, Kenya, Ethiopia, Eritrea, South Sudan, Sudan, and Egypt (Adams, 2007).

According to United Nations Framework convention on climate change (UNFCCC), the impact of climate change on water resources will be more severe in Africa as the continent is likely to face increasing water scarcity with a subsequent potential increase of water conflict across the region due the fact that almost all the 50 river basins in Africa are connected across boundaries (UNFCCC, 2007) in the short term. The societal capacity to adapt to this climate change is low due to less technological and financial resources, which are likely to be exacerbated with the climate change (Byrd and DeMates, 2014). The continent will consequently struggle to produce the required electricity that the people may access through hydropower generation and this is affirmed by the fact that less than 10% of the considerable hydropower potential has been tapped so far (IHA, 2015). This might be partly due to the fact that infrastructures that are resilient to climate change were not considered during the planning stages of the hydropower generation. In the developed countries (Far North of the United States), however, climate change will likely cause rivers to be thick and have a great advantage on hydropower generation (Cherry et al., 2017).

The energy supply sector is by far one of the greatest contributors to global warming due to its enormous emission of greenhouse gases (GHG). In this regard, it becomes important to evaluate the contributions, strategies, risks, uncertainties, and opportunities related to hydropower generation and climate change. The International Hydropower Association elaborates on this relationship through "mitigation, GHG footprints, resilience and adaptation services." According to this association, hydropower as a renewable source of energy is a mitigation strategy even though it is important to understand, predict, and mitigate the potential of GHG footprints of hydropower in specific locations. Nonetheless, this source of renewable energy has not been well exploited in Africa to help mitigate climate change. Africa does not significantly contribute to greenhouse gas emissions, which exacerbate the effects of climate change, but the continent is at a greater advantage in dealing with the effects. In this regard, the continent continues to suffer from droughts and other dry spells that obviously will affect reliable sources of water for hydropower generation due to the changing climate.

The seasonal fluctuations of water resources in Africa (floods and droughts) (Balek, 2011) seem to render hydropower generation of electricity unattractive in the continent. This is because energy generation may only stabilize when water volumes are adequate enough to support the operation (Gerbens-Leenes et al., 2009; Chiang et al., 2013). In a continent mostly hit by frequent droughts, over reliance on hydropower may result in unreliable electricity supply, which the continent is currently facing due to inadequate required technical and financial resources to adapt to the climate change. This explains why even though Africa has a higher hydraulic potential than North/Central America and Europe, their potential exploitation is higher than in Africa (Fig. 1).

3. CLIMATE CHANGE AND PATTERNS OF SOCIETAL CONSUMPTION OF GENERATED HYDROPOWER

3.1 Availability of Electrical Energy Supply in Sub Saharan Africa

In assessing the impacts of expansions of small hydropower plants on climate change mitigation strategies, Kelly-Richards et al. (2017) acknowledged four

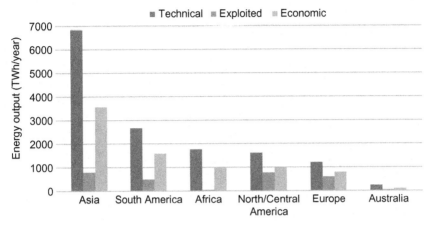

FIG. 1 Hydraulic potential in the world. *(Modified from Pérez-Sánchez, M., et al., 2017. Energy recovery in existing water networks: towards greater sustainability. Water 9 (2), 97.)*

concerns: (1) confusion in small hydropower definitions is convoluting scholarship and policy making; (2) there is a lack of knowledge and acknowledgement of small hydropower's social, environmental, and cumulative impacts; (3) small hydropower's promotion as a climate mitigation strategy can negatively affect local communities, posing contradictions for climate change policy; and (4) institutional analysis is needed to facilitate renewable energy integration with existing environmental laws to ensure sustainable energy development. They recommend that small hydropower development should be regulated by international standard, which should seek to integrate the mitigation and adaptation sides of climate change policy.

Increase in population growth has resulted in increase in energy consumption worldwide. SSA has only 340 million out of its 970 million population having access to electricity (IEA, 2015). However, electricity varies largely across SSA with a 15% rural coverage against a global average of 70% (World Bank, 2015). Hydropower electrification has both economic and climatic implications: it is a tool to tackle energy poverty and green house gas (GHG) emission reduction (Alloisio et al., 2017). In order to sustain the current energy demand within societies, Pérez-Sánchez et al. (2017) reported that strategies related to energy recovery must be promoted and implemented. One such strategy relies on the quantification of the potential hydropower energy recovery in any water system (Gilron, 2014).

The well-endowed river basins in Sub-Saharan Africa, if properly managed in the face of climate change, could provide the needed hydropower energy in the continent. Even though the continent is expanding and is facing population growth, the current generation of hydropower (Table 1) is far below the potential (Fig. 1) reducing consumption drastically.

TABLE 1 World Hydropower Installed Capacity and Generation by Continent in 2014

Continent	Installed Hydropower Capacity (MW) (Excluding Pumped Storage)		Generation (TWh) (Including Pumped Storage for Power Generation)	
	2014	2015	2014	2015
Africa	27,029	30,111	112	116
Central and South Asia	155,731	167,850	455	473
East Asia and Pacific	383,127	443,207	1330	1417
Europe	166,054	218,404	633	599
North America	175,604	199,857	690	680
South America	147,880	152,813	680	684

Data from International Hydropower Association (IHA), 2015. Hydropower Status Report, International Hydropower Association, London and International Hydropower Association (IHA) 2016. Hydropower Status Report, International Hydropower Association, London reports, London, UK.

With regard to the fact that nearly all things depend on electricity to function properly in the 21st century (Sternberg, 2010), it is a great challenge to provide the needed electrical energy to support their functions in Africa. Climate change in one way or the other has also increased the demand for electricity consumption. Extreme temperatures as a result of climate change require electricity for cooling and heating. It is expected that the demand will continue to rise as the years progress (Fig. 2).

Since renewable source of energy cannot be depleted, they could provide reliable and sustainable energy, which has less impact on the climate (Dincer, 2000) to meet the rising needs of power in the society. According to the 2016 report by the International Hydropower Association, the continent that has the lowest power consumption per capita is Africa (181 kWh/year) with an installed capacity of 30 GW as of 2015. Even though the hydropower potentials of Europe and Africa are similar (Fig. 1), Europe has power consumption per capita of 6500 kWh/year with an installed capacity of 218 GW as of 2015. Majority of the people in Africa (645 million) do not have access to electricity due to the low exploitation of the high hydropower potentials that are available.

The deficit in installed hydropower capacity and energy generation on the continent, according to the 2016 report of the International Hydropower Association, has caused the African Development Bank to initiate a new deal on

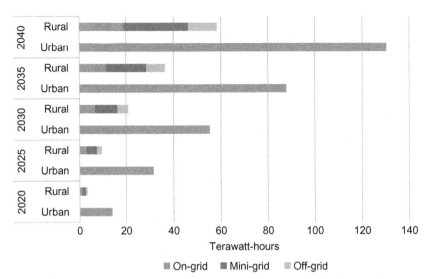

FIG. 2 Projections for electricity demand in Sub-Saharan Africa. *(Based on OECD, 2016. Projected electricity demand from the sub-Saharan African population gaining access to electricity, 2020–40. In: African Economic Outlook 2016, OECD. https://doi.org/10.1787/aeo-2016-graph95-en.)*

energy for Africa. The deal is aimed at achieving worldwide access to energy on the continent by 2025. To achieve the set target of 130 million more on-grid connections in this initiative, hydropower generation will play a major role. An addition of 130 million grid connections, according to the report, will imply 160% increase, which will amount to 160 GW of newly installed capacity on the continent by 2025 (IHA, 2016).

4. CLIMATE CHANGE AND SOCIETAL CHANGE IN THE FACE OF HYDROPOWER ENERGY GENERATION

The society and environment are direct recipients of the effects of climate change (Uzzell, 2008). The ability to adapt to the climate change is a function of social, economic, and political influences and precedence (Barnett and Adger, 2007). Strong and adaptive institutions influence the ability of a society to cope with change so that systems of health and water may function as expected (Handmer et al., 1999). In Sub-Saharan Africa where most institutions mandated to manage the effects of climate change are under resourced and weak due to poverty, warfare, institutional collapse, the least climate variability may result in devastating effects mostly on food and water (Byrd and DeMates, 2014).

Population growth and expansion of urban cities have increased the demand for water and energy, and this is a strong driving force for societies to change (Guy and Marvin, 1996). The expected rapid population growth in Africa, with projection of about 3 billion by 2050 (UN, 2014), suggests increased in healthcare delivery and education in the region. This is because changes in climate will affect the health of the people in a society (Patz et al., 2005) and reliable, affordable, and sustainable electrical energy becomes essential for improving healthcare delivery and interventions for saving lives (Franco et al., 2017). Nevertheless, the current state of healthcare and education in Sub-Saharan Africa, according to (van Der Wat, 2013), does not meet international standards partly due to shortage in electrical capacity. And this was confirmed by Adair-Rohani et al. (2013) who reported that 1 in 4 medical facilities in Africa had no access to electricity while only about one-third of hospitals had access to reliable electricity. In this regard, the continent could take a great advantage of the hydropower potential to improve electrical power generation, which will lead to improved healthcare delivery and education in Africa. Development in hydropower generation will mean that medical facilities have the required energy use for water, temperature control, lighting, ventilation, and clinical processes (WHO, 2017). Consequently, maternal delivery will greatly improve, as well as clinical and emergency procedures in healthcare facilities (mostly in the rural communities) so that most patients' mortality could be avoided since most deaths normally occur due to unstable electrical power generation in the continent (GEA, 2012). In the educational sector, development in hydropower potential will increase advanced research and provide reliable source of electricity for teaching and learning without regular interruptions to studies (Uys et al., 2004).

Since hydropower energy generation is considered a mitigating effect of climate change (IHA, 2015; Kaygusuz, 2016), societies should endeavor to rely on its energy generation opportunities and capacities by finding technological ways of "unearthing water flow from rainwater" and "revealing the complex social relations involved in rainwater management" (Moss, 2000). In response to changing demands and contextual circumstances of the risk and opportunities posed by climate change, advanced societies in developed countries have engineered technical structures to meet the physical expansion of cities and the rising standards of health and environmental protection (Moss, 2000). Most of these advanced societies have expanded and ungraded their physical networks and have acquired state-of-the art technical plants in order to meet an anticipated growth in the use of water resources for generating hydropower energy and a demand for higher performance standards (Auer and Keil, 2012). Berga (2016) reported that the beginning of the 21st century has seen an overwhelming increase in hydropower generation especially in the advanced societies from about 1900 TWh in 1980 to 3600 TWh in 2014. In effect, he continued, much of the hydropower in these advanced societies has been exploited to meet demand.

Africa has a great opportunity to exploit and develop their large hydropower potential through public-private partnership (IHA, 2016). The report by Berga (2016) stated that the effects of climate change on hydropower potential are expected to be small, or even positive in some regions across the globe, even though there will be some variations in some regions. Africa is expected to have an estimated change of 0% in hydropower generation due to climate change by 2050, according the report. The rise in investment in the generation of hydropower in the region (van Der Wat, 2013) is therefore a step in the right direction since the youth, mostly young women, are continuously depending on electricity for their businesses to thrive. Most of these young women who previously relied solely on agriculture for their livelihoods are now into businesses that need increased reliable supply of electrical energy for the survival of such businesses (Bryceson, 2002). This implies that, as the hydropower potential is tapped to develop the energy sector, majority of the youth in the society will be empowered through energy security and supply to grow Africa's economy thereby attaining the Millennium development goals of the continent and contributing less to climate change.

5. CONCLUSION

Sub-Saharan Africa has a vast potential of very little exploited hydropower. The population in the region continues to rise requiring increase in energy consumption. The current energy consumption in the region is the lowest worldwide, leaving a great majority of the population without electricity. The society in response to the deficit in hydropower exploitation remains underdeveloped in many sectors since energy is a major necessity for things to function in order to bring about development in a society in the 21st century.

Climate change will continue to affect and influence water discharge courses worldwide, and Africa as a developing society, will subsequently suffer severely from these effects (both negative and positive) even though the continent contributes insignificantly to this change. The inability of the continent to adapt to these climate changes has resulted in various social menace (frequent droughts, poverty, diseases, famine, conflicts, etc.) across the region.

Presently, for Africa to benefit enormously from hydropower energy generation, the region will have to employ the technologies employed in advanced societies to harness hydropower potentials. Fortunately, climate change invariably will have little impact (0% change) on hydropower generation in Africa by 2050, which implies the continent can comfortably depend on it for energy generation. Hydropower energy exploitation is therefore a great prospect for Africa to provide reliable, affordable, and environment friendly electrical energy to reach over 75% of the population (645 million) living without electricity in Sub-Saharan Africa.

REFERENCES

Adair-Rohani, H., et al., 2013. Limited electricity access in health facilities of sub-Saharan Africa: a systematic review of data on electricity access, sources, and reliability. Glob. Health Sci. Pract. 1 (2), 249–261.

Adams, O., 2007. The quest for cooperation in the Nile water conflicts: a case for Eritrea. Afr. Sociol. Rev. 11 (1), 95–105.

Adger, W.N., et al., 2003. Adaptation to climate change in the developing world. Prog. Dev. Stud. 3 (3), 179–195.

Adger, W.N., et al., 2009. Are there social limits to adaptation to climate change? Clim. Chang. 93 (3), 335–354.

African Development Bank, 2015. Sustainable Cities and Structural Transformation. African Economic Outlook. Retrieved from https://www.afdb.org/fileadmin/uploads/afdb/Documents/Publications/AR2015/Annual_Report_2015_EN_-_Chapter_2.pdf.

Alloisio, I., et al., 2017. Energy poverty alleviation and its consequences on climate change mitigation and African economic development. A policy brief published by Fondazione Eni Enrico Mattei (FEEM) publications. Retrieved from http://feem.it.

Auer, J., Keil, J., 2012. State-of-the-Art Electricity Storage Systems: Indispensable Elements of the Energy Revolution (March 8, 2012). Current Issues. Deutsche Bank Research. Available at: https://ssrn.com/abstract=2882202.

Balek, J., 2011. Hydrology and Water Resources in Tropical Africa. Elsevier, New York. 208p.

Barnett, J., Adger, W.N., 2007. Climate change, human security and violent conflict. Polit. Geogr. 26 (6), 639–655.

Berga, L., 2016. The role of hydropower in climate change mitigation and adaptation: a review. Engineering 2 (3), 313–318.

Bryceson, D.F., 2002. Multiplex livelihoods in rural Africa: recasting the terms and conditions of gainful employment. J. Mod. Afr. Stud. 40 (1), 1–28.

Byrd, R., DeMates, L., 2014. 5 Reasons why climate change is a social issue, not just an environmental one. HUFFPOST. http://www.huffingtonpost.com/rosaly-byrd/climate-change-is-a-socia_b_5939186.html. (Accessed 16 May 2017).

Cherry, J.E., et al., 2017. Planning for climate change impacts on hydropower in the far north. Hydrol. Earth Syst. Sci. 21 (1), 133–151.

Chiang, J.-L., et al., 2013. Potential impact of climate change on hydropower generation in southern Taiwan. Energy Procedia 40, 34–37.

Cobbinah, P.B., et al., 2015. Africa's urbanisation: implications for sustainable development. Cities 47, 62–72.

van Der Wat, S., 2013. Hydro in Africa: navigating a continent of untapped potential. Hydroworld 21, 1–21.

Dincer, I., 2000. Renewable energy and sustainable development: a crucial review. Renew. Sust. Energ. Rev. 4 (2), 157–175.

Franco, A., et al., 2017. A review of sustainable energy access and technologies for healthcare facilities in the global south. Sustain. Energy Technol. Assess. 22, 92–105.

GEA, 2012. Global Energy Assessment—Toward a Sustainable Future. Cambridge University Press and the International Institute for Applied Systems Analysis, Cambridge/Laxenburg.

Gerbens-Leenes, P., et al., 2009. The water footprint of energy from biomass: a quantitative assessment and consequences of an increasing share of bio-energy in energy supply. Ecol. Econ. 68 (4), 1052–1060.

Gilron, J., 2014. Water-energy nexus: matching sources and uses. Clean Techn. Environ. Policy 16 (8), 1471–1479.

Guy, S., Marvin, S., 1996. Disconnected policy: the shaping of local energy management. Eviron. Plann. C. Gov. Policy 14 (1), 145–158.

Handmer, J.W., et al., 1999. Societal vulnerability to climate change and variability. Mitig. Adapt. Strateg. Glob. Chang. 4 (3), 267–281.

IEA, 2012. World Energy Outlook, International Energy Agency Report 2012, Paris. Retrieved from. http://worldenergyoutlook.org.

IEA, 2015. Africa Energy Outlook. World Energy Outlook Report. OECD/International Energy Agency, France.

IHA, 2015. Hydropower Status Report. International Hydropower Association, London.

IHA, 2016. Hydropower Status Report. International Hydropower Association, London.

Kaygusuz, K., 2016. Hydropower as clean and renewable energy source for electricity generation. J. Eng. Res. Appl. Sci. 5 (1), 359–369.

Kelly-Richards, S., Silber-Coats, N., Crootof, A., Tecklin, D., Bauer, C., 2017. Governing the transition to renewable energy: a review of impacts and policy issues in the small hydropower boom. Energy Policy 101, 251–264.

Moss, T., 2000. Unearthing water flows, uncovering social relations: introducing new waste water Technologies in Berlin. J. Urban Technol. 7 (1), 63–84.

OECD, 2016. Projected electricity demand from the sub-Saharan African population gaining access to electricity, 2020–40. In: African Economic Outlook 2016. OECD, https://doi.org/10.1787/aeo-2016-graph95-en.

Patz, J.A., et al., 2005. Impact of regional climate change on human health. Nature 438 (7066), 310–317.

Pérez-Sánchez, M., et al., 2017. Energy recovery in existing water networks: towards greater sustainability. Water 9 (2), 97.

Shove, E., 2010. Beyond the ABC: climate change policy and theories of social change. Environ. Plan. A 42 (6), 1273–1285.

Sternberg, R., 2010. Hydropower's future, the environment, and global electricity systems. Renew. Sust. Energ. Rev. 14 (2), 713–723.

UN, 2014. World Urbanization Prospects: The 2014 Revision, Highlights. Department of Economic and Social Affairs, Population Division, United Nations.

UNFCCC, 2007. Impacts, Vulnerabilities and Adaptation in Developing Countries. United Nations Framework Convention on Climate Change (UNFCCC), Germany.

Uys, P.M., et al., 2004. Technological innovation and management strategies for higher education in Africa: harmonizing reality and idealism. Educ. Media Int. 41 (1), 67–80.

Uzzell, D., 2008. Challenging assumptions in the psychology of climate change. InPsych 30 (4), 10.

WHO, 2017. Health and sustainable development: energy access and resilience. http://www.who.int/sustainable-development/health-sector/health-risks/energy-access/en/.

World Bank, 2015. Sustainable Energy for All. Database Retrieved from: http://data.worldbank.org.

World Energy Resources, 2015. Charting the Upsurge in Hydropower Development. Retrieved from: https://www.worldenergy.org/wp-content/uploads/2015/05/World-Energy-Resources_Charting-the-Upsurge-in-Hydropower-Development_2015_Report2.pdf.

Chapter 6

Renewable Energy and Sustainable Development

Samuel Gyamfi, Nana S.A. Derkyi, Emmanuel Y. Asuamah, Israel J.A. Aduako
University of Energy and Natural Resources (UENR), Sunyani, Ghana

Chapter Outline

1. Introduction	75	5.3 Technical Potential	83
2. Geographic and Climatic		5.4 Economic Potential	83
Conditions of West Africa	76	5.5 Solar Photovoltaic (PV)	
3. Renewable Energy, a		and Concentrated Solar	
Necessity for Sustainable		Power	83
Development	78	6. Wind Energy Potentials in	
4. Renewable Energy and		West Africa	85
Sustainable Development	80	7. Biomass for Energy Potential	
4.1 Connection Between		in West Africa	85
Renewable Energy and		8. Barriers to Renewable	
Sustainable Development	81	Energy Integration	88
4.2 Sustainability Indicators for		9. Observations and Discussions	89
Renewable Energy	82	10. Policy Recommendations	
5. Renewable Energy Potential		and Way Forward	90
in West Africa	82	11. Conclusion	91
5.1 Theoretical Potential	82	References	92
5.2 Geographic Potential	82	Further Reading	93

1. INTRODUCTION

The notion of sustainable development can be expounded in many ways, but at its core, it is an approach to development that seeks to balance various, and often competing, needs against an awareness of the environmental, social, and economic limitations we face as a society. In other words, it delivers basic environmental, social, and economic services without threatening the viability of natural, built, and social systems upon which these services depend (Bugaje, 2006). Sustainable development has been captured in the policies and plans

Sustainable Hydropower in West Africa. https://doi.org/10.1016/B978-0-12-813016-2.00006-X

of many African countries. Energy consumption is considered as a vital indicator in sustainable development.

The major global energy challenges are securing energy supply to meet growing demand, providing everybody with access to energy services, and curbing energy's contribution to climate change (Moomaw et al., 2011). The threatening fact is that more than 620 million people in sub-Saharan Africa (two-thirds of the population) live without electricity (IEA, n.d.-a). This is the part of the world with greater availability of renewable energy (RE) resources (Cabeza et al., 2009) and yet lack access to electricity. For developing countries in West Africa, energy is needed to stimulate production, income generation, and social development, and to reduce the serious health problems caused by the use of fuel wood, charcoal, dung, and agricultural waste.

Even though hydropower has been harnessed and has proven to be of great benefit in West Africa, there are a lot more RE sources like Solar power, Wind power, and Biofuels that could also be harnessed for electricity generation. For example, the average solar radiation is estimated to be between 5 and 6 kWh/m^2/day across the West African Subregion (Economic Commission for Africa, 2002). This gives the region an opportunity to integrate solar energy into its generation mix. Wind and biofuel on the other hand present themselves with high potentials on the West African subregion (Ambali et al., 2011).

In a world of increasing demand for energy and finiteness of fossil-fuel resources and its accompanied environmental concerns and for the fact that Sustainable development cannot be achieved if we continue to threaten the viability of natural environment, there is the need for the West African Subregion as a matter of urgency to integrate clean and reliable energy resources into their generation mix. This chapter presents potential of RE resources in West Africa based on theoretical and geographic potential. The analysis is based on the solar maps (direct/global irradiation), wind speed, and land cover maps of West Africa. The connection between RE and sustainable development has also been drawn in the chapter.

2. GEOGRAPHIC AND CLIMATIC CONDITIONS OF WEST AFRICA

West Africa lies between latitudes 4°N and 28°N and longitudes 15°E and 16°W. The Gulf of Guinea is the southern boundary, while that to the north is Mauritania, Mali, and Niger; the Mount Cameroon/Adamawa Highlands and the Atlantic Ocean constitute the eastern and the western limits. West Africa includes 16 countries: Benin, Cape Verde, Gambia, Ghana, Guinea, Guinea-Bissau, Ivory Coast, Liberia, Mali, Mauritania, Niger, Nigeria, Senegal, Sierra Leone, Togo, and Burkina Faso. West Africa has an area of 6 million km^2 covering one-fifth of Africa (West Africa, n.d.).

West Africa has wet and dry seasons resulting from the interaction of two migrating air masses. The first is the hot, dry tropical continental air mass of the northern high-pressure system, which gives rise to the dry, dusty, Harmattan winds that blow from the Sahara over most of West Africa from November to February; the maximum southern extension of this air mass occurs in January between latitudes 5° and 7°N. The second is the moisture-laden, tropical maritime or equatorial air mass that produces southwest winds. The maximum northern penetration of this wet air mass is in July between latitudes 18° and 21°N. Where these two air masses meet is a belt of variable width and stability called the Intertropical Convergence Zone (ITCZ).

The north and south migration of this ITCZ, which follows the apparent movement of the sun, controls the climate of the region. The lowland climates of West Africa are characterized by uniformly high sunshine and high temperatures throughout the year. Mean annual temperatures are usually above 18°C. Areas within 10° of the equator have a mean annual temperature of about 26°C with a range of 1.7–2.8°C; the diurnal range is 5.6–8.3°C. Between latitudes 10°N and the southern part of the Sahara mean monthly temperatures can rise to 30°C, but the annual range is 9°C and diurnal range 14–17°C. In the central Sahara, temperatures in the shade in July may be as high as 58°C during the day and as low as 4°C at night; mean annual temperature ranges from 10°C to 35°C. Fig. 1 shows the map of West Africa.

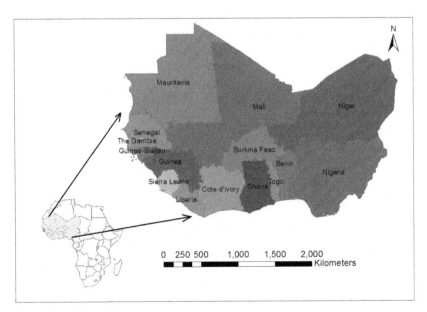

FIG. 1 Map of West Africa.

3. RENEWABLE ENERGY, A NECESSITY FOR SUSTAINABLE DEVELOPMENT

The Sustainable Development Goals (SDGs) embrace the need for economic development that gives everyone a fair chance of leading a decent life. The 7th and the 13th goals acknowledge the importance of "affordable, reliable, sustainable, modern energy for all and climate change mitigation and the reduction of environmental and health impacts" (West Africa, n.d.). Sustainable development on the other hand is the development that meets the needs of the present without compromising the ability of future generations to meet their own needs (Brundtland, 1987). Sustainable development could also be seen as the urgency of making progress toward economic development that could be sustained without depleting natural resources or harming the environment (United Nations Commission on Sustainable Development, 2007). These outlined goals seek to address issues of the environment. There is a big relationship between sustainable development and energy and hence the need to look into the sources of our energy (Lund, 2007). The concepts of sustainable development as explained here all point to one direction and that is protecting the natural environment for the future generation while meeting our immediate needs.

Our current sources of energy in West Africa and the world at large come from fossil fuels. These resources contribute immensely to the destruction of the natural environment and thus cannot be seen as sources of energy for sustainable development. Apart from the emissions associated with fossil fuels, they are limited and if we continue to depend on them, the future generation would have to look for their sources of energy supply and this does not conform to the concept of sustainable development (Bach, 1981). Fig. 2 shows the energy mix of some selected West African countries compared to that of the world.

From Fig. 2, there is a limited diversification of energy sources; countries depend basically on oil and natural gas. Hydro plays an important role in Ghana but the percentage of oil is still high (>60%). Benin, Ghana, and Togo

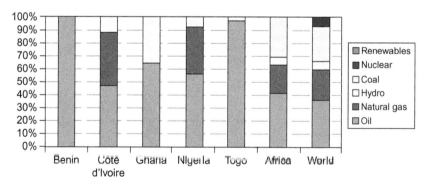

FIG. 2 Commercial energy consumption mix of some selected West African countries in 2005 compared to that of the world (EIA, 2005).

are dependent on imported oil. Africa and the world at large depend so much on fossil-fuel resources. The use of these fuels contributes to the destruction of the environment and so for a sustainable development to be achieved, there is the need to have a diversified source of energy. Because sustainable development does not destroy the natural environment and for the fact that fossil fuels destroy the environment means that they cannot be considered on the sustainable development agenda. Amidst the environmental impacts of these fuels, their lifespan is limited and so the future generation would have to cater for their own sources of energy, which is in contrast to the concept of sustainable development.

RE resources on the other hand do not pollute the natural environment and they are inexhaustible, which makes them the best candidates for sustainable development (World Bank, 2007; Adenikinju, 2008). Fig. 3 shows the installed electricity generation mix of some selected West African countries.

It shows that two inputs account for over 95% of total generation capacity of all the countries. This high concentration provides little energy security for the countries as the disruptions in one or both of these sources would have a significant impact on electricity generations in these countries (Adenikinju, 2008).

Climate change mitigation has got to do with reducing or tackling the factors that contribute to climate change. Fossil fuel as a source of energy contributes immensely to the destruction of the environment (Holling, 1997; N. R. Council, 2010). This is among some of the reasons why we should rally behind RE development. RE is projected to play a central role in most GHG mitigation strategies (West Africa, n.d.; Edenhofer et al., 2012). Access to modern energy services, whether from renewable or nonrenewable sources, is closely correlated with measures of development, particularly for those countries at early development stages. The link between adequate energy services and achievement of the Millennium Development Goals (MDGs) was defined explicitly in the Johannesburg Plan of Implementation that emerged from the World Summit on Sustainable Development in 2002 (IEA, n.d.-b). As emphasized by a number

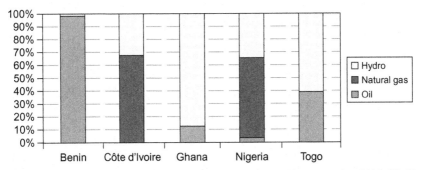

FIG. 3 Installed generation capacity of some selected West African countries, 2005 (World Bank, 2007).

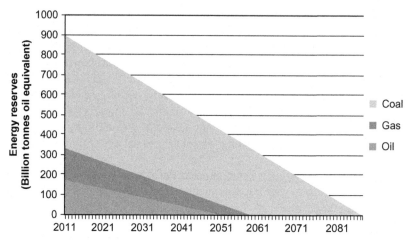

FIG. 4 Chart showing static range of the different fossil fuels (Hoffert, 2010).

of studies, providing access to modern energy (such as electricity or natural gas) for the poorest in society is crucial for the achievement of any of the eight MDGs (Meadows, 1998b; Creutzig and Kammen, 2009).

There is clear evidence that fossil fuels are not sustainable, and it is expected that oil, gas, and coal will be depleted by 2051, 2061, and 2090, respectively (Fig. 4). The three most important factors in selecting new energy resources that ensure sustainability and supply security are that they must be renewable, locally available at reasonable costs, and environmentally friendly (Kazem, 2011).

SDGs placed an emphasis on modern energy for all, but our current source of energy clearly cannot provide energy for all as enshrined in the SDGs. Therefore, the only hope for the future of West African subregion and the world at large is to tap into the inexhaustible source of energy we have. The West African Subregion has a great RE potential that could be harnessed to help curb the energy crisis we face today.

4. RENEWABLE ENERGY AND SUSTAINABLE DEVELOPMENT

Various factors such as limitation of fossil-fuel resources, negative impacts on environment, fossil-fuel prices, political/social disputes, and their effects on supplying sustainable energy are among the reasons that have compelled nations to move toward the development of a modern structure to secure supply of energy, environmental protection, and efficiency improvement of energy systems with economic gains. Hence, most countries have begun to realize that the need for sustainability in energy production and consumption is significantly vital. Fig. 5 shows the scheme for sustainable development as the convergence of three sustainability pillars.

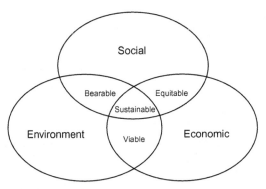

FIG. 5   Sustainable development scheme.

Historically, energy continues to be the pivot of economic and social development of all countries around the world. Though it has brought great economic prosperity, the way it is produced and used has adversely affected local, regional, and global environments, hence the ongoing debate about making energy systems more sustainable. This debate has largely centered on detailed discussions around the different energy sources and their likely emissions during thermochemical and biological operations. Support has been expressed for RE systems, owing to their more favorable environmental qualities.

4.1   Connection Between Renewable Energy and Sustainable Development

The connection between RE and SD can be viewed as a hierarchy of goals and constraints that involve both global and regional or local considerations. Mitigation of anthropogenic climate change will be one of the strong driving forces behind increased use of RE technologies worldwide. RE plays a central role in most GHG mitigation strategies that must be technically feasible and economically efficient. Knowledge about technological capabilities and models for optimal mitigation pathways is therefore important. However, energy technologies, economic costs and benefits, and energy policies depend on the societies and natural environment within which they are embedded. Sustainability challenges and solutions crucially depend on the geographic setting (solar radiation), socioeconomic conditions (inducing energy demand), inequalities within and across societies, fragmented institutions, and existing infrastructure (electric grids) (Evans et al., 2009; Hermann et al., 2014).

Maximizing the environmental benefits often depends on the specific technology, management, and site characteristics associated with each RE project, especially with respect to land use change (LUC) impacts. Life cycle assessments for electricity generation indicate that GHG emissions from RE technologies are, in general, considerably lower than those associated with fossil-fuel options. The maximum estimate for concentrating solar power (CSP) and wind energy is less than or equal to $100\,g\,CO_2eq/kWh$, and median values for all RE range from 4 to $46\,g\,CO_2eq/kWh$ (Edenhofer et al., 2012).

4.2 Sustainability Indicators for Renewable Energy

Efforts to ensure SD can impose additional constraints or selection criteria on some climate change mitigation pathways and may compel policymakers and citizens to accept tradeoffs. For each additional boundary condition placed on an energy system, some development pathways are eliminated as being unsustainable, and some technically feasible scenarios for climate mitigation may not be viable if SD matters (Edenhofer et al., 2012).

Sustainability indicators for energy are important for monitoring progress made in energy subsystems consistent with sustainability principles, and its measurement and reporting can have a pervasive effect on decision making (Kearney et al., 2002; Singh et al., 2015). Measuring energy sustainability raises either technical and conceptual issues (Ohimain, 2010) or updated methodologies (Creutzig and Kammen, 2009). Recently, some progress has been made toward developing a uniform set of energy indicators for SD (taking into account the economy, society, and environment components) (Prakash and Bhat, 2010). Quantitative indicators for RE technologies include generated electricity price, GHG emissions, availability of renewable sources, the efficiency of energy conversion, land requirements, and water consumption (Prakash and Bhat, 2010).Other approaches compare the different RE subsystems based upon their performance, net energy requirements, GHG emissions, and other indicators (Evans et al., 2009).

5. RENEWABLE ENERGY POTENTIAL IN WEST AFRICA

The potential of RE resources to produce energy can be viewed in different dimensions. When investigating the potential of RE, it is necessary to distinguish between categories of potentials that require series of processing steps and assumptions. The following gives the various categories of RE potentials.

5.1 Theoretical Potential

The theoretical potential has to do with the amount of resource available without considering any conversion efficiencies and losses. It is the maximum amount of energy that is physically available from a certain source. In the case of solar energy, this would be equal to the total solar radiation impinging on the evaluated surface (Hermann et al., 2014).

5.2 Geographic Potential

The geographic potential may be seen as an intermediate step toward calculating the technical potential of RE resource. The geographic potential takes into account areas that are suitable and usable for specific RE employment.

Depending on the details of available geographic data, an appropriate set of exclusion criteria can be set to realistically estimate the available land area (e.g., exclusion of urban areas for large-scale wind power production, protected land, sloped areas, and water bodies) (Hermann et al., 2014).

5.3 Technical Potential

The technical potential is the geographic potential minus the losses from conversion into secondary energies and constrained by the requirements related to large-scale installation (e.g., spacing factors representing spacing and servicing areas of solar power plants or wind turbines, as well as (grid-) transportation losses). Technological, structural, ecological, and legislative restrictions and requirements are accounted for. In other words, RE technical potential represents the achievable energy generation of a particular technology given system performance, topographic limitations, environmental, and land-use constraints (Hermann et al., 2014; Kearney et al., 2002).

5.4 Economic Potential

Economic potential is the proportion of the technical potential that can be utilized economically. It takes into account costs and other socioeconomic factors (e.g., fuel and electricity prices, other opportunity costs, and land prices).

5.5 Solar Photovoltaic (PV) and Concentrated Solar Power

Photovoltaic Solar Energy is nowadays a mature technology for production of electricity with low-carbon emissions and for a wide range of applications. West Africa is located under Sunbelt (ECREEE, 2012).

According to the Sunbelt Potential of Photovoltaics (Kearney et al., 2002), in an ambitious scenario with adequate political support and deployment measurements, PV energy could be a sustainable and competitive energetic technology providing up to 12% of the Sunbelt countries' electricity demand by 2030 (ECREEE, 2012). Ghana, for instance, has a monthly average solar irradiation measured by the Mechanical Engineering Department of Kwame Nkrumah University of Science and Technology (KNUST) to be between 4.4 and 5.6 kWh/m^2/day. There is a large regional variations with 5.3 kWh/m^2/day in Kumasi in the cloudy semideciduous forest region to 7.7 kWh/m^2/day at Wa in the dry savannah region. Northern Ghana, including northern parts of Brong-Ahafo and the Volta, has a monthly average of 4.0–6.5 kWh/m^2/day, which is very high (Singh et al., 2015). Table 1 summarizes the daily average solar irradiation of all the West African countries.

Concentrated Solar Power (CSP) involves power generation using direct normal solar radiation. The global radiation consists of diffuse and direct and unlike the Solar PV technology that makes use of both the diffuse and the direct

TABLE 1 Solar Irradiance of West African Countries (NREL, 2005)

Country	Solar Resources (kWh/m²/day)
Benin	4.89–5.68
Cape Verde	5.71–6.12
Gambia	5.60–5.74
Ghana	4.02–6.07
Guinea	5.02–5.66
Guinea-Bissau	5.46–5.57
Ivory Coast	4.63–5.61
Liberia	4.62–4.98
Mali	5.64–6.45
Mauritania	5.65–6.68
Niger	5.69–6.73
Nigeria	3.96–6.29
Senegal	5.49–5.97
Sierra Leone	4.74–5.34
Togo	5.00–5.55
Burkina Faso	5.57–6.15

radiations, the concentrated Solar Power can only utilize the direct radiations from the sun. Adequate solar resources of sites vary from 1800 to 2800 kWh/m², allowing from 2000 to 6500 annual full-load operating hours with the solar element, depending on the available radiation at the particular site and the relative size of solar field aperture, heat storage capacity, and nominal power of the power block (ECREEE, 2012). CSP may provide distributed and centralized solutions for electricity supply, and it is one of the main candidate technologies to find a viable transition to a sustainable electricity supply, catalyst for sustainable development. Another reason to invest in CSP for sustainable is the possibility of combined generation of electricity and heat to achieve the highest possible efficiencies for energy conversion. In addition to electricity, such plants can provide steam for absorption chillers, industrial process heat, or thermal seawater desalination. West Africa has all the solar resources to make this possible. Fig. 6 shows the solar resource map of Africa. It shows the potential of solar energy in the West Africa subregion.

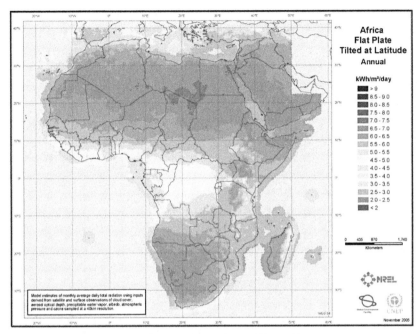

FIG. 6 Solar resource map of Africa (Solar Map of West Africa, n.d.)

6. WIND ENERGY POTENTIALS IN WEST AFRICA

For wind energy, the calculation of the technical potential is complex due to the fact that the resources differ considerably over the area of an entire country. Nevertheless, an approximation can be made using the annual average wind speed of a region, as there is an established relationship between the wind speed and the average full-load hours (or the capacity factor) of a wind turbine (Hoogwijk and Graus, 2008). The theoretical and geographical potential of wind energy is presented in Table 2.

Fig. 7 shows the wind speed resource in West Africa and Africa at large. It outlines the potential of wind energy measured at 80 m.

7. BIOMASS FOR ENERGY POTENTIAL IN WEST AFRICA

Lignocellulose is globally recognized as the preferred biomass for the production of a variety of fuels and sustainable chemicals and fuels industry with significant benefits in agricultural development. Lignocellulose represents the most widespread and abundant source of carbon in nature and is the only source that could provide a sufficient amount of feedstock to satisfy the world's energy and chemicals needs in a renewable manner (Lynd et al., 2003). The most dominant parameters used to determine the potential of energy crops are land

TABLE 2 Wind Speeds of West African Countries 50 m Height (NREL, 2005)

Country	Wind Resources (m/s^2)
Benin	1.3–4.1
Cape Verde	5.4–6.7
Gambia	3.3–3.7
Ghana	2.2–3.5
Guinea	2.0–3.4
Guinea-Bissau	3.1–3.8
Ivory Coast	1.9–3.4
Liberia	1.9–3.4
Mali	2.1–4.6
Mauritania	2.2–7.5
Niger	1.9–4.6
Nigeria	2.0–3.5
Senegal	1.6–4.6
Sierra Leone	2.2–3.3
Togo	2.3–3.2
Burkina Faso	1.4–2.7

and yield (Stecher et al., 2013). The West African subregion has a vast land size that is fertile for any crop production and for that matter energy crop for biofuel production. Table 3 shows the potential of biofuel in some selected countries in West Africa.

Nigeria for instance produces 134 million liters of ethanol per annum from five major commercial-scale ethanol distilleries (Ohimain, 2010). Biodiesel production projects that have been initiated include Biodiesel Nigeria Limited in Lagos State, Aura Bio-Corporation in Cross River State, and the Shashwat Jatropha in Kebbi State (Ohimain, 2010).

Ghana introduced a bioenergy policy in 2010, which was created to substitute the country's petroleum oil with 10% biofuels by 2020 and 20% by 2030, respectively. The country introduced this policy with the intention to utilize the vast majority of biomass resources that are abundant in Ghana for the generation of transport fuels and electricity. Biomass resources, such as cassava, sugarcane, maize, and jatropha oil seeds, were identified as potential feedstocks for bioethanol and biodiesel production in Ghana (Oteng-Adjei, 2010).

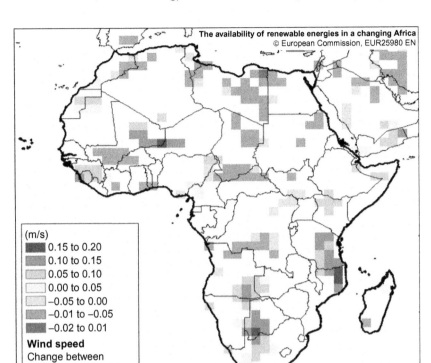

FIG. 7 Wind resource map of Africa (European Commission, 2016).

TABLE 3 Biofuel Potential in Selected West African Countries in Megaliters (ML) (Sekoai and Yoro, 2016)

Country	Raw Material	Biodiesel (ML)	Ethanol (ML)
Benin	Cassava	–	20
Ghana	Jatropha	50	–
Ivory Coast	Molasses	–	20
Mali	Molasses	–	20
Niger	Jatropha	10	–
Nigeria	Sugarcane	–	70
Senegal	Molasses	–	15
Togo	Jatropha	10	–
Burkina Faso	Sugarcane	–	20

8. BARRIERS TO RENEWABLE ENERGY INTEGRATION

Barriers to RE integration for sustainable development have to do with circumstances or obstacles that keep or slow down the utilization of RE technologies.

Energy security: there is no commonly accepted definition of the term "energy security" and its meaning is highly context dependent (European Commission, 2016; Sekoai and Yoro, 2016). At a general level, it can best be understood as robustness against (sudden) disruptions of energy supply (IEA, 2006). Thinking broadly across energy systems, one can distinguish between different aspects of security that operate at varying temporal and geographical scales (Bazilian and Roques, 2008). Two broad themes can be identified that are relevant to energy security, whether for current systems or for the planning of future RE systems: availability and distribution of resources, and variability and reliability of energy supply. Given the interdependence of economic growth and energy consumption, access to a stable energy supply is a major political concern and a technical and economic challenge facing both developed and developing economies, since prolonged disruptions would create serious economic problems. The introduction of renewable technologies that vary on different time scales, ranging from minutes to seasonal, adds a new concern to energy security. Not only will there be concerns about disruption of supplies by unfriendly agents, but also of the vulnerability of energy supply. When High penetration of RE sources exists, the fluctuating characteristics of these sources (solar and wind power), together with the high variability of the demand curve in rural communities, generate technical challenges in terms of creating a grid with a high-quality power supply.

Financing RE is another factor that impedes the smooth integration of RE technologies into the current energy mix. The cost of investing in RE is very high and sometimes it becomes very difficult getting help from the financial institutions. Developing new renewable resources will require large initial investments to build infrastructure. These investments increase the cost of providing renewable electricity, especially during early stages (Beck and Martinot, 2004).

The following scale-up cost of RE technology utilization:

Developers must find publicly acceptable sites with good resources and with access to transmission lines. Potential wind sites can require several years of monitoring to determine whether they are suitable and all these come with a cost. The other unseen part is the time value of money, which is the time taken to check if a particular site would really be feasible.

Permitting issues for conventional energy technologies are generally well understood, and the process and standards for review are well defined. In contrast, renewables often involve new types of issues and ecosystem impacts. And standards are still in the process of development.

Marketing: In the past, individuals had no choices about the sources of their electricity. But electricity deregulation has opened the market so that customers have a variety of choices. Start-up companies must communicate the benefits of renewables to customers in order to persuade them to switch from traditional sources. Public education will be a critical part of a fully functioning market if renewables are to succeed.

Installation, operation, and maintenance: Workers must be trained to install, operate, and maintain new technologies, as well as to grow and transport biomass fuels. Some renewables need operating experience in regional climate conditions before performance can be optimized.

Another factor that increases the cost of RE technology is the fact that compared with renewables, nuclear and fossil-fuel technologies enjoy a considerable advantage in government subsidies for research and development. In addition to receiving subsidies for research and development, conventional generating technologies have a lower tax burden. Fuel expenditures can be deducted from taxable income, but few renewables benefit from this deduction since most do not use market-supplied fuels. Income and property taxes are higher for renewables, which require large capital investments but have low fuel and operating expenses. (N. R. Council, 2010; Edenhofer et al., 2012)

9. OBSERVATIONS AND DISCUSSIONS

The West African subregion presents an array of varied energy mix, resources, and policies. What is common, however, is the availability of abundant RE resources in almost every country. Nigeria is a case of an abundance of both conventional and renewable resources but with very poor infrastructural support to harness the renewable resources (Bugaje, 2006). RE resources, abundant in all the West African countries, would provide a major breakthrough in finding a solution to its energy crisis. Energy is indeed needed for economic growth, and lack of adequate energy services is certainly a constraint to sustainable development. It limits the potentials of meeting basic needs of those who need energy to undertake essential domestic, agricultural, and educational tasks; to support health and transport services, and to initiate or develop manufacturing or trading enterprises as stated in the SDGs (Meadows, 1998b).

A major area of concern raised by some writers on sustainable development is the many barriers to development in West Africa and Africa as a whole such as abject poverty, the poor state of health and education, and environmental degradation (Krantz, 2001). However, the solution to these barriers lies to a large extent in addressing the energy challenges facing the continent and foremost as reliable and sustainable energy supply forms the very basis for solving the other socioeconomic problems (Krantz, 2001). Most of the West African countries are engaging in ways to support Renewable Energy Technologies (RETs); Mali is one such country to take a step to provide a case study of urgency in addressing sustainable energy policy especially in view of the environmental degradation associated with the traditional energy use patterns (Panwar et al., 2011). Ghana has taken a step by enacting a Renewable energy Act 2011, which seeks to increase the generation mix of renewable to 10% and also providing a feed-in tariffs for renewables (Government of Ghana, 2011). From the potential of RE resources seen in West Africa, it is clear that renewables could be a great deal in achieving the SDGs.

10. POLICY RECOMMENDATIONS AND WAY FORWARD

Imperfections and distortions in the market coupled with unfavorable financial, institutional, and regulatory environments imply that governmental intervention is not only desirable but also a must to promote RETs. The role of governments in technology transfer has been outlined in the IPCC special report on technology transfer, which is relevant for renewables too (IPCC, 2000). The role includes generic actions to remove barriers, building human and institutional capacity, setting up research and development infrastructure, creating an enabling environment for investment, and providing information and mechanisms to promote RETs (Union of Concerned Scientists, 1999). Some measures that the governments in the West African countries could take to help smoothen the RET integration in their current generation mix to meet the increasing energy demand for sustainable development include restructuring their current energy sector. Governments within the West African subregion should introduce competition and remove other controls such as those creating separate entities for generation and distribution in the electricity sector, allow private sector entry and diluting or removing controls on energy pricing, fuel use, fuel import, and capacity expansion. Institutional measures such as setting up independent regulatory bodies may be needed for the success of these policy actions. The basic purpose of this is to increase the efficiency of the energy sector through facilitating market competition. The initial impact of such measures may be unfavorable to RETs due to increased competitiveness. However, in the long term, a liberalized energy market may provide a better environment for the healthy growth of RETs. (Painuly, 2001)

Policies should be put in place to guarantee a market for RETs. Since RE is not able to compete in the energy market with existing barriers, energy suppliers may be required by law to include a part of the energy from renewables in their supply mix. Such measures include Renewable Energy Act 2011 in Ghana, which seeks to provide a policy framework to support the development and utilization of RE sources and to create an enabling environment to attract investment in RE sources (Government of Ghana, 2011). Non-Fossil Fuel Obligation (NFFO) law in the United Kingdom and Electricity Feed Law (EFL) in Germany guarantee predetermined electricity prices for competitively selected RE projects. It promotes reduced cost of RETs due to a competitive process for project selection. An example of such law is shown in Table 4, which is the feed-in tariff gazetted for Ghana in November 2014.

Economic/financial incentives could be a way forward to improve the utilization of RETs in West Africa. Governments should provide capital subsidies

TABLE 4 Feed-In Tariff (FIT) Ghana Gazette—November 2014 (Effective October 1, 2014) (Government of Ghana, 2011)

Generation Source	FIT (Ghp/kWh)	FIT (US$/kWh)	Maximum Capacity (MW)
Wind with grid stability systems	55.7369	17.4254	300
Wind without grid stability systems	51.4334	16.08	30
Solar PV with grid stability/storage systems	64.4109	20.1372	150
Solar PV without grid stability	58.3629	18.2464	150
Hydro ≤10 MW	53.6223	16.7643	No limit
Hydro (>10 MW and ≤100 MW)	53.8884	16.8475	No limit
Biomass	56.0075	17.51	No limit
Biomass (enhanced technology)	59.035	18.4565	No limit
Biomass (plantation as feed stock)	63.2891	19.7865	No limit

for installation of RE Systems. Tax exemption, credit facilities, and third-party financing mechanisms are other measures to overcome the barriers related to RETs. Stakeholders can also be educated and supplied with the necessary tools to evaluate the RETs and design implementation. The campaigns are both general in nature and targeting specific RET product promotion (Painuly, 2001).

11. CONCLUSION

West Africa is facing a serious energy crisis. This is not due to lack of energy resources, but rather the poor state of infrastructural support and appropriate technology to harness these resources, especially the renewable ones. Now that the world's attention is focused more on energy sources that would be able to sustain us for the future and also safeguard the environment from further degradation, West Africa should be able to take an opportunity in that regard since it has abundant renewable resources. This chapter presents a comprehensive overview of the potentials of RETs such as solar PV, Solar thermal, and Wind Energy based on theoretical and geographical data taken from NREL database. The review indicates that the location of West Africa points to the fact that it is endowed with most renewable energy resources, the subregion is located under the Sunbelt. Mauritania has as high as 6.68 kWh/m^2 of solar radiation and 7.5 m/s wind speed, which runs throughout the country and most of the countries within the subregion have a similar experience. Most of the countries in West Africa have taken steps to address the ever-increasing energy demand through the sustainable ways, thus through engagement in the support for RETs.

REFERENCES

Adenikinju, A., 2008. West Africa Energy Security. University of Ibadan Center for Energy Economics at the University of Texas at Austin Kumasi Institute of Energy, Technology and Environment.

Ambali, A., Chirwa, P.W., Chamdimba, O., van Zyl, W.H., 2011. A review of sustainable development of bioenergy in Africa: an outlook for the future bioenergy industry. Sci. Res. Essays 6 (8), 1697–1708.

Bach, W., 1981. Fossil fuel resources and their impacts on environment and climate. Int. J. Hydrog. Energy 6 (2), 185–201.

Bazilian, M., Roques, F. (Eds.), 2008. Analytical Methods for Energy Diversity and Security Elsevier Global Energy.

Beck, F., Martinot, E., 2004. Renewable energy policies and barriers. Encycl. Energy 34 (3), 365–383.

Brundtland, G.H., 1987. Our common future: report of the World Commission on Environment and Development. Med. Confl. Surviv. 4 (1), 300.

Bugaje, I.M., 2006. Renewable energy for sustainable development in Africa: a review. 10, 603–612.

Cabeza, L.F., Hollands, T., Jäger-waldau, A., Khattab, N., Tamaura, Y., Xu, H., Zilles, R., Meier, A., Morofsky, E., Mujumdar, A., 2009. Direct Solar Energy. pp. 1–113. (Chapter 3).

Creutzig, F., Kammen, D., 2009. The Post-Copenhagen roadmap towards sustainability: differentiated geographic approaches, integrated over goals. Innovations, 301–321.

Economic Commission for Africa, 2002. Harnessing Energy Resources for Sustainable Development in Africa.

ECREEE, "Renewable Energy in Western Africa: Situation, Experiences and Tendencies". 2012.

Edenhofer, O., Pichs Madruga, R., Sokona, Y., 2012. Renewable energy sources and climate change mitigation. 6 (4). (Special Report of the Intergovernmental Panel on Climate Change).

EIA, 2005. Environmental Impact Assessment. 2004 (20).

European Commission, 2016. Wind speed map of Africa. Available from: https://www.google.com.gh/search?q=wind+speedmap+of+west+africa. (Accessed 27 November 2016).

Evans, A., Strezov, V., Evans, T.J., 2009. Assessment of sustainability indicators for renewable energy technologies. Renew. Sust. Energ. Rev. 13 (5), 1082–1088.

Government of Ghana, 2011. Renewable Energy Act 2011: Act 832. pp. 1–27.

S. Hermann, A. Miketa, and N. Fichaux, "Estimating the renewable energy potential in Africa, IRENA-KTH Working Paper," 2014.

Hoffert, M.I., 2010. Farewell to fossil fuels? Science 329 (5997), 1292–1294.

Holling, C.S., 1997. Regional Responses to Global Change. Conserv. Ecol. 1 (2), https://doi.org/10.5751/CE-00023-010203.

Hoogwijk, M., Graus, W., 2008. Global potential of renewable energy sources: a literature assessment. EcoFys (March), 45.

IEA, 2006. World energy outlook 2006. In: Outlook. pp. 600.

IEA, n.d.-a. Africa focus. Available from: http://www.worldenergyoutlook.org/resources/energydevelopment/africafocus/. (Accessed 11 October 2016).

IEA, n.d.-b, "Technology Roadmap—Biofuels for Transport, Paris: IEA." [Online]. Available from: http://www.reee.sacities.net/.

IPCC, 2000 In: Metz, B,, Davidson, O., Martens, J.-W., Van Rooijen, S., Van Wie Mcgrory, L. (Eds.), Methodological and Technological Issues in Technology Transfer.

Kazem, H.A., 2011. Renewable energy in Oman: status and future prospects. Renew. Sust. Energ. Rev. 15 (8), 3465–3469.

Kearney, A.T., Hauff, J., Verdonck, M., Derveaux, H., Dumarest, L., Alberich, J., Malherbe, J.-C., 2002. unLOCKING THE Sunbelt Potential of Photovoltaics. 22 (3), 179–189.

L. Krantz, "The sustainable livelihood approach to poverty reduction," Division for Policy and Socio-Economic Analysis, February, p. 44, 2001.

Lund, H., 2007. Renewable energy strategies for sustainable development. Energy 32 (6), 912–919.

Lynd, L., Blottnitz, H.V., Tait, B., Boer, J.D., Pretorius, I., Rumbold, K., Zyl, W.V., 2003. Converting plant biomass to fuels and commodity chemicals in South Africa: a third chapter? S. Afr. J. Sci. 99 (11–12), 499–507.

D. Meadows, "Indicators and information systems for sustainable," p. 78, 1998b.

Moomaw, W., Yamba, F., Kamimoto, M., Maurice, L., Nyboer, J., Urama, K., Weir, T., 2011. In: Introduction: renewable energy and climate change. IPCC Spec. Rep. Renew. Energy Sources Clim. Chang. Mitig. pp. 161–208.

N. R. Council, 2010. Hidden Costs of Energy. National Academies Press, Washington, DC.

NREL, "RETScreen 4, Clean energy project analysis. 2005.

Ohimain, E.I., 2010. Emerging bio-ethanol projects in Nigeria: their opportunities and challenges. Energ. Policy 38 (11), 7161–7168.

Oteng-Adjei, J., 2010. Draft bioenergy policy of Ghana. In: pp. 1–29. Energy Commission, Accra.

Painuly, J.P., 2001. Barriers to renewable energy penetration: a framework for analysis. Renew. Energy 24 (1), 73–89.

Panwar, N.L., Kaushik, S.C., Kothari, S., 2011. Role of renewable energy sources in environmental protection: a review. Renew. Sust. Energ. Rev. 15 (3), 1513–1524.

Prakash, R., Bhat, I.K., 2010. A figure of merit for evaluating sustainability of renewable energy systems. Renew. Sust. Energ. Rev. 14 (6), 1640–1643.

Sekoai, P., Yoro, K., 2016. Biofuel development initiatives in sub-Saharan Africa: opportunities and challenges. Climate 4 (2), 33.

G. Singh, Gauri Singh, Safiatou Alzouma Nouhou and Mohamed Youba Sokona, "Ghana Renewables Readiness Assessment," No. November, 2015.

Solar Map of West Africa. n.d. Available from: https://www.google.com.gh/search?q=solar+map+of+west+africa (Accessed 27 November 2016).

K. Stecher, A. Brosowski, and D. Thrän, "Biomass potential in Africa," IRENA, p. 44, 2013.

Union of Concerned Scientists, 1999. Barriers to renewable energy technologies. In: Powerful Solutions: Seven Ways to Switch America to Renewable Electricity. Available from: http://www.ucsusa.org/clean_energy/smart-energy-solutions/increase-renewables/barriers-to-renewable-energy.html (Accessed 10 November 2016).

United Nations Commission on Sustainable Development, 2007. Framing sustainable development The Brundtland Report—20 years on. p. 2.

West Africa, n.d. "Integrating crops and livestock in West Africa." [Online]. Available from: http://www.fao.org/docrep/004/x6543e/x6543e01.htm (Accessed 31 October 2016).

World Bank, 2007. World Development Indicators. World Bank, Washington, DC.

FURTHER READING

Bazilian, M., Hobbs, B.F., Blyth, W., MacGill, I., Howells, M., 2011. Interactions between energy security and climate change: a focus on developing countries. Energ. Policy 39 (6), 3750–3756.

Dartanto, T., 2013. Reducing fuel subsidies and the implication on fiscal balance and poverty in Indonesia: a simulation analysis. Energ. Policy 58, 117–134.

Meadows, D., 1998a. In: Indicators and information systems for sustainable development. A Rep. to Balat. Gr. pp. 1–25.

Sgobba, V., Nocera, D., Guldi, D., 2009. 2009 renewable energy issue. Chem. Soc. Rev. 38 (1), 165–184.

Verbruggen, A., Fischedick, M., Moomaw, W., Weir, T., Nadaï, A., Nilsson, L.J., Nyboer, J., Sathaye, J., 2010. Renewable energy costs, potentials, barriers: conceptual issues. Energ. Policy 38 (2), 850–861.

Chapter 7

The Potential and the Economics of Hydropower Investment in West Africa

Samuel Gyamfi, Nana S.A. Derkyi, Emmanuel Y. Asuamah
University of Energy and Natural Resources (UENR), Sunyani, Ghana

Chapter Outline

1. **Introduction**   95
   1.1. Layout of West Africa   97
2. **Hydropower Potential in West Africa**   97
   2.1. Niger River   98
   2.2. Senegal River   98
   2.3. Volta River   99
3. **Climate Change Uncertainties and the Prospects of Hydropower in West Africa**   99
4. **Economics of Hydropower in West Africa**   101
   4.1. Cost of Hydropower Project 102
5. **Economic Benefits of Hydroelectric Power Generation Over Other RE Sources**   103
   5.1. The Base Load Power Case   104
6. **Factors Affecting Private Sector Investment**   104
7. **Observations and Discussions**   105
8. **Conclusions**   105
   References   106
   Further Reading   107

1. INTRODUCTION

Over the years, West Africa has experienced a rising demand for electricity supply. This increasing demand for power has been as a result of increasing economic activities, urbanization, and population growth (Keong, 2005). Hydropower is able to provide cheap and continuous access to power supply thereby helping to alleviate poverty in most West African countries since economic development is tied to the availability of energy supply (Carley et al., 2011). Energy consumption is a very crucial factor for every aspect of our life and as population increases, the demand for energy obviously will increase (Bildirici and Gökmenoğlu, 2016). Apart from the fact that population increase has a significant effect on the demand for energy, as people aspire for more comfort and as their needs increase, they find ways to make life easy thus

Sustainable Hydropower in West Africa. https://doi.org/10.1016/B978-0-12-813016-2.00007-1

increasing their energy consumption (Keong, 2005; Stern, 2011). For example, when people decide to wash their clothes and dishes with electrical appliances, they require energy and that increases their demand for energy. The problem we have now is that even though our demand for energy keeps on increasing, the sources that we get the energy from are depleting. Most of our energy sources today come from fossil fuels that are finite and poses threat to the natural environment due to the amount of greenhouse gases they release into the environment (Höök and Tang, 2013; Bartle, 2002).

Hydropower has been used for hundreds of years because it is often cost competitive as an energy source and can be utilized with relatively basic technology. Today, hydropower remains cost competitive in many regions, but it is earning extra attention due to a growing emphasis on renewable energy (RE) sources and the increase in overall demand for energy (Kalitsi, 2003; Lee and Leal, 2014). West African countries have enormous potential in terms of hydropower generation. Burkina Faso for instance has economically feasible hydropotential of 216 GWh/year. This potential runs through the rest of the West African countries. Ghana and Nigeria are West African countries with major hydropower development with installed capacities of 1938 and 1072 MW, respectively (Lee and Leal, 2014).

West Africa, which consists of 18 countries and covers about 17% of the African continent (CEDEAO-CSAO/OCDE, n.d.; Speth, 2010), is covered by 71% of 28 border river basins with the important ones being the Niger, the Senegal, the Volta, the Lake Chad, and the Comoe (West Africa Eco Zone, n.d.) (Fig. 1).

There is a large potential for hydropower in West Africa; hence, this chapter seeks to assess the potential while considering the economics involved in investing into this power generation sources compared to other renewable sources.

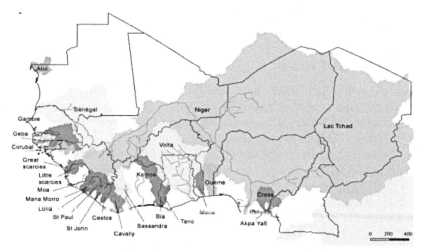

FIG. 1   Water resources of West Africa (CEDEAO-CSAO/OCDE, n.d.).

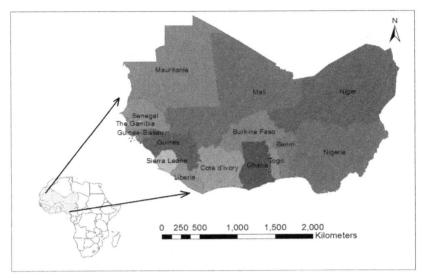

FIG. 2 West Africa Eco Zone (Edenhofer et al., 2012).

This chapter is organized into various sections as follows: Section 1 provides a general outlook of the energy issues in some West African countries and the contribution of hydropower to meeting energy demands. Section 2 takes a look at the potential of small or micro hydropower in West Africa. Section 3 provides some analysis of the economics involved in investing in hydropower in West Africa compared to other renewable sources. Section 4 concludes the paper.

1.1. Layout of West Africa

West Africa is the westernmost subregion of Africa. It occupies an area in excess of 6,140,000 km^2, or approximately one-fifth of Africa (Fig. 2). The vast majority of this land is plains lying below 300 m above sea level, though isolated high points exist in numerous states along the southern shore (IPCC, 2011). The northern section of West Africa is composed of semiarid terrain known as Sahel, a transitional zone between the Sahara and the savannahs of the western Sudan. Forests form a belt between the savannas and the southern coast, ranging from 160 to 240 km in width (IPCC, 2011).

2. HYDROPOWER POTENTIAL IN WEST AFRICA

The annual global technical potential for hydropower globally is estimated at 14,576 TWh with the corresponding estimated total capacity potential of 3721 GW. The undeveloped capacity ranges from 47% in Europe up to about 92% in Africa (IPCC, 2011). Thus, Africa has a large hydroresource that is yet to be developed. Table 1 shows regional estimated technical and installed capacity.

TABLE 1 Regional Estimated Technical Capacity and Installed Capacity in GW (Edenhofer et al., 2012).

Regions	Estimated Capacity	Installed Capacity	Undeveloped Potential (%)
North America	388	153	61
Latin America	608	159	74
Europe	338	179	47
Africa	283	23	92
Asia	2037	402	80
Australasia	67	13	81

The climates of West Africa range from desert and savannah to lowland woodlands and tropical rainforests. In the less arid areas, the inland population centers tend to be confined to the banks of a few significant rivers. One of the world's major rivers, the Congo is nearly 3000-mile long, making it the second-largest river in Africa. With depths measured at >750 ft, the Congo is the deepest river in the world. The Congo runs much faster than most other rivers of such size and is second only to the Amazon in terms of the amount of water it discharges at its mouth (Paventi, 2014). Apart from the Congo River, there are a lot more of such rivers that could be harnessed to generate hydropower. Examples of such rivers are as follows:

2.1. Niger River

Niger is the longest river that flows its entire length within West Africa. At almost 2600 miles, the Niger flows through five countries in the region before emptying into the Atlantic Ocean at the Gulf of Guinea. Since the late 20th century, Niger's delta has become a major source of oil and natural gas, as well as a major transportation source for the entire region (Paventi, 2014).

2.2. Senegal River

Rising in the mountains of Guinea, the Senegal River flows through Mauritania and the country that shares its name before spilling into the Atlantic Ocean at Saint-Louis. Hydroelectricity and agriculture are the primary economic contributions made to the region by the 1700-mile river, which is the second longest in West Africa (Paventi, 2014).

2.3. Volta River

Formed by the confluence of the Black, White, and Red Voltas, the Volta River is a relatively shallow watercourse that flows through Burkina Faso and Ghana before reaching its mouth at the Gulf of Guinea on Africa's Atlantic coast. The Akosombo Dam in Ghana, a major source of hydroelectric power in the region, created Lake Volta, which is the world's largest reservoir (Paventi, 2014).

Nigeria has a gross theoretical hydropower potential of 42,750 GWh/year and a technically feasible potential of 32,450 GWh/year both evaluated in 1975 and 1980. The technically feasible potential, evaluated in 1980, is 28,800 GWh/year (Abdul-Aziz, 2015). So far, Nigeria has developed 26% of the technically feasible potential. As at 2015, 2040 MW of hydrocapacity was in operation in the country (Abdul-Aziz, 2015). Hydroproduction in 2014 was about 5345 GWh, of a total of 30,126 GWh (NCC, 2014). The Energy Commission of Nigeria (ECN) has estimated the total capacity of the country's hydropower potential at 11,500 and 3500 MW for large and small hydropower systems, respectively.

Mali places great importance on the development of their hydroresources, as a foundation for their future progress. It has up to 800 MW of new hydrocapacity planned in the long term, with several medium-sized projects at the feasibility stage. Detailed financial studies have been conducted for four of them (Kalitsi, 2003).

Apart from the already existing hydropower facilities, Ghana still has a 2000-MW of additional untapped hydropower resource of which 1205 MW is expected to emanate from large hydroplants and the remaining from medium and small hydroplants (Kabo-Bah et al., 2016).

The West African Subregion places itself as the most endowed subregion with high potential for hydropower generation in Africa. Liberia, which has suffered a long-term civil war, has had most of its power facilities damaged or destroyed, and the Liberia Electricity Corporation regards development of the country's large hydropotential as a major hope for the basis of economic recovery (Kabo-Bah et al., 2016).

Table 2 provides a summary of hydropower potential in some countries in West Africa. Guinea, where the 75-MW Garafiri project recently began operation, reports plans to develop more hydropower, starting with the 104-MW Kaleta scheme (IJHD, 2005).

3. CLIMATE CHANGE UNCERTAINTIES AND THE PROSPECTS OF HYDROPOWER IN WEST AFRICA

Despite the overwhelming prospects of hydropower in West Africa, the impacts of climate change on this notable energy resource cannot be overlooked. This arises from the fact that climate change involves long-term changes in

TABLE 2 Hydropower Potential in West Africa Countries in GWh/year (IJHD, 2005)

Country	Theoretical Potential	Technical Potential	Economic Potential
Benin	1676	<1676	–
Ghana	–	10,600	–
Guinea	26,000	19,400	14,500
Guinea-Bissau	–	300	300
Ivory Coast	46,000	>12,400	12,400
Liberia	–	11,000	–
Mali	–	>5000	–
Niger	–	1300	1300
Nigeria	42,750	32,450	29,800
Senegal	–	4250	2050
Sierra Leone	–	6800	–
Togo	–	1700	–
Burkina Faso	>600	–	216

precipitation and temperature patterns, which inevitably pose threat to the sustainability of hydropower resources. An underpinning factor to this long-term climatic variation is the influence of land cover and land use change (Feddema et al., 2005; Gordon et al., 2003). The aforementioned factors have significant impact on the water balance of hydropower resources in terms of the runoffs into dams, evaporation from reservoirs, precipitation into the reservoir, and eventually the water levels in the dam (Kabo-Bah et al., 2016). For instance, Ghana's perennial energy crisis has been partly attributed to the recurrent low water levels in the Akosombo Dam (Fiagbe and Obeng, 2006; Mul et al., 2015; Gyau-Boakye, 2001). It is projected by Obahoundje et al. (2017) that in the next 25 years, energy generation from Bui Hydropower Plant will reduce by 46% and 2.5%, respectively, under climate change dry condition (15% reduction in rainfall) and current conditions. Notwithstanding this challenge, there still exist high prospects for the hydropower industry in West Africa since measures could be implemented to adapt and mitigate climate change impacts on hydropower generation. Considering the prevailing climatic conditions, special attention must be given to the impacts of climate change when making a detailed assessment of the economic feasibility of hydropower projects in West Africa.

4. ECONOMICS OF HYDROPOWER IN WEST AFRICA

The economics of a hydropower plant is quite different from that of any other type of power plant since various considerations such as water supply, irrigation, and river navigation are involved besides regular economic aspects of cost of generated power. In fact, some of these aspects, such as effect on irrigation or recreation facilities, are difficult to quantify. Hence, true economic analysis of a hydropower plant, especially a large hydropower plant, is a mix of quantitative and qualitative approaches.

Cost can be measured in a number of ways, and each way of accounting for the cost of power generation brings its own insights. The costs that can be examined include equipment costs, replacement costs, financing costs, total installed cost, fixed and variable operating and maintenance costs (O&M), fuel costs, and the levelized cost of energy (LCOE). The analysis of costs can be very detailed, but for purposes of comparison and transparency, the approach used here is a simplified one. This allows greater scrutiny of the underlying data and assumptions, improved transparency and confidence in the analysis, as well as facilitating the comparison of costs by country or region for the same technologies in order to identify what are the key drivers in any differences (Feddema et al., 2005).

The major benefits and cost components for estimating annual net benefits from a hydropower plant are defined as follows (Gordon et al., 2003).

1. Gross power benefits: These benefits reflect the income from sale of power or avoided cost of power if the hydropower plant did not exist and power has to be taken from costlier source.
2. Benefits of avoided pollution: Relative to alternative types of power generation, such as a coal-fired plant, hydropower production generates less air pollution or greenhouse gases. The avoided pollution is considered as a benefit of hydropower projects.
3. Costs of operation: This type of costs reflects investment costs for the project, anticipated future reinvestment costs, and current operation and maintenance (O&M) costs.
4. Benefits of project services: Beyond power generation, hydroelectric projects may offer benefits such as flood control, water supply, irrigation, river navigability, and improvement of infrastructure and economic prosperity of the region.
5. Costs of environmental measures: Many licensing decisions introduce operating conditions designed to protect, mitigate damages to, or improve environmental quality. These changes may result in direct costs and/or reduced power values from the viewpoint of the hydropower station owner. There are direct costs associated with, for example, construction of fish passage facilities. Similarly, due to environmental measures to protect flora and fauna, sometimes flow of water is restricted that may reduce power generation either because they cause direct losses in available flow or they shift power

generation from periods when energy prices are high to periods when energy prices are low.

6. Benefits of environmental measures: Environmental measures, such as fish screens or changes in minimum flow requirements, can improve fish and wildlife resources, recreational opportunities, and other aspects of environmental quality. Since these benefits are different from the direct revenue from sale of power, they are often referred to as "nonpower" benefits.

4.1. Cost of Hydropower Project

4.1.1 Investment Cost

RE projects are particularly sensitive to high financing costs, as they are characterized by high upfront costs and low operational costs. The initial costs of hydropower plants are usually found to vary between 1000 and 5000 Euros per kW depending on the size of country and location. In the case of Ghana, a 400-MW hydropower plant constructed (Bui Power Plant) 11 years ago had an initial cost of $622 million ($1.555 million/MW). The total cost of the Felou hydroelectric of Mali was US$125.00 million (Oji et al., 2016). In case of greenfield projects, where no dam has existed before the hydropower project, civil engineering works typically account for 65%–75% of the total initial cost and meeting the environmental and legal criteria requires 15%–20%. The cost of plant machinery, such as turbine, generator, and control systems, may account for only 10% of the total initial cost (Oji et al., 2016).

The LCOE of different RE generation technologies using both the social discount rate and the true financial cost for investors operating in Ghana and other countries has been calculated (Fiagbe and Obeng, 2006). As per the true financial costs, a low-cost scenario and a high-cost scenario were considered. Social discount rates measure "the rate at which a society is willing to trade present for future consumption" (Mul et al., 2015). Policymakers use them in cost-benefit analyses of social projects. Countries with high social discount rates will tend to favor projects with short-run benefits as opposed to those that deliver benefits in the long term. The World Bank typically uses a social discount rate of 10% to assess infrastructure investments in developing countries. This rate, twice as high as the one used in OECD countries, reflects a higher time discount in poorer countries. Cost-benefit analyses in Ghana use a social discount rate of 12% in Ghanaian policy documents (Gyau-Boakye, 2001). Three elements determine financing costs: the debt-to-equity ratio, the cost of equity, and the cost of debt. The cost of debt depends on the interest rate, maturity, and grace period of the loans provided. Equity investors usually require rates of return of at least twice the cost of debt, as they assume a higher risk. Projects with high equity shares therefore bear higher financing costs. Smaller and riskier projects typically require higher equity shares as they struggle to be attractive. Data collected for Ghana showed a debt-equity ratio of 70:30, which is similar to that observed in developed countries with lower perceived risks (Fiagbe and Obeng, 2006).

Project developers in West Africa can access both commercial and concessional finance. Commercial finance is faster to obtain, but it charges highest rates and typically offers lower maturities. Concessional finance offers better terms but usually involves larger transaction costs and a slow turnaround. Both domestic and international banks can provide commercial debt for hydropower project developers. International banks usually offer better rates and leaner processes, as they have more experience in RE and more capital available (Fiagbe and Obeng, 2006).

Access to international finance is key for the viability of RE projects in Ghana, given the high cost of local finance. Average debt rates for international finance, with some concessional finance, are 7.5%. Fully commercial debt would require 12%–16% interest rates. Domestic lending rates are significantly higher, at 21%–37%. Loan maturities are 9–15 years. Rates of return required by equity investors can be as high as 30% (Obahoundje et al., 2017). An additional factor to take into account is the emergence of China as a nontraditional financier in the region. China is now the largest financier of power projects in Africa (IRENA, 2012). Chinese investment is mainly focused on large hydroprojects.

Hydropower projects qualify as Clean Development Mechanism (CDM) project under the Kyoto Protocol. Out of the 2062 projects registered by the CDM Executive Board by 1 March 2010, 562 are hydropower projects (IPCC, 2011). Most of these projects were from Asia and South America. With about 92% of Africa's hydropower potential yet to be developed, this flexible mechanism provides unique opportunity for developing the potential hydrosites for power generation.

5. ECONOMIC BENEFITS OF HYDROELECTRIC POWER GENERATION OVER OTHER RE SOURCES

Apart from the fact that hydropower plants are cheaper than other RE sources, they have other benefits that cannot be overlooked. For instance, the Bui power plant in Ghana has an additional irrigation benefit of high-yield crops on 30,000 ha of fertile land in an economic Free Zone (Cordaro, 2008). Hydroelectric power plants with accumulation reservoirs offer incomparable operational flexibility, since they can immediately respond to fluctuations in the demand for electricity. The flexibility and the storage capacity of hydroelectric power plants make them more efficient and economical in supporting the use of intermittent sources of RE, such as solar energy or wind energy (Energy Commission of Ghana, 2015). The other renewable sources of energy like solar and wind are intermittent; thus, the rate at which the resources vary is very high and so whenever power is been generated from those sources, there is always the need to have a backup system from conventional sources or hydropower.

5.1. The Base Load Power Case

Base load is the minimum level of demand on an electrical supply system over 24 h. Base load power sources are those plants that can generate dependable power to consistently meet demand. They are the foundation of a sound electrical system (Energy Commission of Ghana, 2015). Large hydroplants are the future for generation mix. They can operate and serve the base loads required. The Akosombo hydroelectric dam of Ghana serves as the base load plant for the country. Similarly, Kainji, Shiroro, and Jebba of Nigeria also serve the country's base load (Abdul-Aziz, 2015).

6. FACTORS AFFECTING PRIVATE SECTOR INVESTMENT

The factors affecting private sector investment in hydropower include financial impediments, regulatory impediments, and capacity impediments. Financial impediments include the high costs inherent in the energy sector, including project preparation, tenders, and importing commodities such as crude oil and gas. Limited access to funding is also a problem. Most countries have poor or nonexistent sovereign credit ratings, limiting their access to international credit markets, and domestic capital markets are narrow. Potentially valuable financial instruments like project bonds are generally not available. Financial risks include insufficient cost recovery, elastic demand, nonpayment or inability to pay for services, and foreign exchange risk. Possible initiatives to overcome these difficulties entail investing in cost-reducing technology, using syndicated loans, expanding pension funds and project bonds, increasing partial risk guarantees, using indexing for foreign currency risk, and investing in prepayment meters (Dambudzo, 2009).

Regulatory impediments involve the lack of independent or impartial regulators in some countries; lack of competition or open access to transmission and distribution networks; one-off power purchase agreements (PPAs) rather than standard PPAs; weak procurement laws; inefficient or nontransparent tendering processes that result in cancelled, postponed, or disputed tenders; poor contract laws; and tariffs that are set by the government with no provision for inflation or changes in cost. The regulatory risks are the breach of contracts, the partiality of regulators, and the inability to raise tariffs to cover costs. Possible initiatives include setting multiyear tariffs with automatic adjustment clauses, unbundling utilities into different components to open up competition, setting renegotiation clauses in original contracts, and outlining performance targets for public authorities and private concessionaires (Dambudzo, 2009).

Capacity impediments include the lack of skills among public officials to manage Public Private Partnerships (PPPs); most local judicial systems do not have the capacity to handle complex contracts or disputes; and regional and subnational regulatory frameworks are not harmonized, which poses problems for projects that cross borders. Capacity risks consist of bureaucratic procedures that effectively halt or delay a project; change in administrations and consequently different rules for investors; uneven policies in different countries; and nationalization or expropriation. Possible initiatives for tackling these impediments include streamlining public agencies to minimize bureaucracy, hiring and developing individuals who have experience in PPPs, and strengthening regional PPP capacity and cooperation (Dambudzo, 2009).

7. OBSERVATIONS AND DISCUSSIONS

The West African Subregion places itself as the most endowed subregion with high potential for hydropower generation in Africa. Africa remains the region with the lowest ratio of deployment-to-potential ratio, and the opportunities for growth are very large (Feddema et al., 2005). West Africa has the potential to generate power from hydro but looking at the potential and the utilization so far, West Africa has much to do as a subregion that wants to develop. Existing data show that most countries use hydro to supply the country's base load demand and this is due to the fact that they are reliable. Most West African countries are tapping into their hydropower potentials. Nigeria has developed 26% of the technically feasible potential. As at 2015, 2040 MW of hydrocapacity was in operation in the country. Mali places great importance on the development of their hydroresources, as a foundation for their future progress. It has up to 800 MW of new hydrocapacity planned in the long term, with several medium-sized projects at the feasibility stage; detailed financial studies have been conducted for four of them (Kalitsi, 2003). Ghana for instance has developed three hydropower plants, namely, Akosombo hydroelectric power station, Bui power, and Kpong hydropower plants. Looking at the potential that the subregion has, development of small hydropower plants could be a head way for the region to develop since development moves with energy demand and energy security issues (Keong, 2005). Initial costs of hydropower plants are usually found to vary between 1000 and 5000 Euros per kW depending on the size of country and location. In the case of Ghana, a 400-MW hydropower plant (Bui Power Plant) 11 years ago had an initial cost of $622 million ($1.555 million/MW) and cost of Felou hydroelectric plant of Mali was US$125.00 million (Oji et al., 2016).

8. CONCLUSIONS

There has been a recent wave of public opposition to hydropower, particularly schemes involving reservoirs impounded by large dams. However, it is clear that hydrostations that are planned, constructed, and operated with adequate consideration of environmental and social aspects have a major role to play in world energy supply in the future. They will be particularly beneficial in meeting the needs of developing countries like nations in West Africa. It is time for nations in West Africa to make good use of the opportunities they have and progress in that respect. West Africa has got potential for other renewable sources but hydroenergy presents itself as the most economically feasible source of power generation. The problem with the other renewable sources has got to do with their intermittency, which necessitates the need for a supporting back-up power. This and many other issues make the economics of hydropower more feasible. Also, hydropower plants could serve other purposes like irrigation. Similar to many other RE technologies, hydropower projects incur high initial capital costs but relatively lower lifetime operating costs and could last for 70–100 years.

REFERENCES

Abdul-Aziz, D., 2015. Hydropower Development in Nigeria: The Beginning. North-South Power Company, Shiroro Hydroelectric Power Station in Niger State, Nigeria.

Bartle, A., 2002. Hydropower potential and development activities. Energ. Policy 30 (14), 1231–1239.

Bildirici, M.E., Gökmenoğlu, S.M., 2016. Environmental pollution, hydropower energy consumption and economic growth: evidence from G7 countries. Renew. Sustain. Energy Rev. (October). https://doi.org/10.1016/j.rser.2016.10.052.

Carley, S., Lawrence, S., Brown, A., Nourafshan, A., Benami, E., 2011. Energy-based economic development. Renew. Sust. Energ. Rev. 15 (1), 282–295.

CEDEAO-CSAO/OCDE, n.d., Les bassins fluviaux transfrontaliers. Atlas régional de l'Intégration en Afrique de l'Ouest, Série Espaces.

Cordaro, M., 2008. Understanding Base Load Power. What It Is and Why It Matters. Affordable Reliable Electricity Alliance, New York, p. 5.

Dambudzo, M., 2009. Increasing Private Investment in African Energy Infrastructure. Ministerial and Expert Roundtable of the NEPADOECD Africa Investment Initiative.

O. Edenhofer, R. Pichs Madruga, and Y. Sokona, Renewable Energy Sources and Climate Change Mitigation (Special Report of the Intergovernmental Panel on Climate Change), vol. 6, no. 4. 2012.

Energy Commission of Ghana, 2015. Energy (supply and demand) outlook for Ghana, p. 44.

Feddema, J.J., Oleson, K.W., Bonan, G.B., Mearns, L.O., Buja, L.E., Meehl, G.A., Washington, W.M., 2005. The importance of land-cover change in simulating future climates. Science 310, 1674–1678.

Fiagbe, A.K.Y., Obeng, D.M., 2006. Optimum operations of hydropower systems in Ghana when Akosombo Dam level is below minimum design level. J. Sci. Technol. 26 (2).

Gordon, L., Dunlop, M., Foran, B., 2003. Land cover change and water vapour flows: learning from Australia. Philos. Trans. R. Soc. Lond. Ser. B Biol. Sci. (358), 1973–1984.

Gyau-Boakye, P., 2001. Environmental impacts of the Akosombo Dam and effects of climate change on the lake levels. Environ. Dev. Sustain. 17–29.

Höök, M., Tang, X., 2013. Depletion of fossil fuels and anthropogenic climate change—a review. Energ. Policy 52, 797–809.

IJHD, 2005. World Atlas & Industry Guide. International Journal of Hydropower and Dams, Wallington, Surrey, 383p.

IPCC, 2011. Renewable energy sources and climate change mitigation special report of the intergovernmental panel on climate change.

IRENA, 2012. Hydropower. Renew. Energy Technol. Cost Anal. Ser. 1 (3/5), 44.

Kabo-Bah, A.T., Diji, C.J., Nokoe, K., Mulugetta, Y., Daniel Obeng-Ofori, D., Akpoti, K., 2016. Multiyear rainfall and temperature trends in the Volta River Basin and their potential impact on hydropower. Climate 4 (4), 49.

Kalitsi, E.A.K., Kalitsi and Associates, 2003. Problems and prospects for hydropower development. Prep. by Work. African Energy Expert. Oper. NGPAD Energy Initiat. Probl. Prospect. Hydropower Dev. Africa, no. June.

Keong, C.Y., 2005. Energy demand, economic growth, and energy efficiency—the Bakun dam-induced sustainable energy policy revisited. Energ. Policy 33, 679–689.

Lee, N.C., Leal, V.M.S., 2014. A review of energy planning practices of members of the Economic Community of West African States. Renew. Sust. Energ. Rev. 31, 202–220.

Mul, M., Sidibé, Y., Annor, F., Ofosu, E., Boateng-Gyimah, M., Ampomah, B., Addo, C., 2015. Balancing hydro generation with sustainable ecosystem management. Water storage and hydro development for Africa. Palais des Congrès la Palmeraie, Marrakesh, pp. 46–50.

NCC, 2014. NCC limited annual report 2014–15.

Obahoundje, S., Ofosu, E.A., Akpoti, K., Kabo-bah, A.T., 2017. Land use and land cover changes under climate uncertainty: modelling the impacts on hydropower production in Western Africa. Hydrology 4, 2.

Oji, C., Soumonni, O., Ojah, K., 2016. Financing renewable energy projects for sustainable economic development in Africa. Energy Procedia 93 (March), 113–119.

Paventi, J., 2014. Rivers in West Africa. Available from: http://www.ehow.com/list_5594861_rivers-west-africa.html%5Cn%5Cn. (Accessed 12 November 2016).

Speth, P., 2010. Impacts of Global Change on the Hydrological Cycle in West and Northwest Africa. Springer, Heidelberg, Dordrecht, London, New York.

Stern, D.I., 2011. The role of energy in economic growth. Ann. N. Y. Acad. Sci. 1219 (1), 26–51.

"West Africa Eco Zone." n.d. Available from: https://www.google.com.gh/search?q=Layout+of+West+Africa.

FURTHER READING

Africa Energy Unit and the World Bank, 2011. Options for the development of Liberia's energy sector. no. 63735, pp. 32–33.

Graphic Online, 2016. The power of Bui Dam.

Sherman, J., 2004. Hydroelectric power. Available from: http://water.usgs.gov/edu/hydroadvantages. html (Accessed 21 November 2016).

Wagner, H.J., Mathur, J., 2011. Introduction to hydro energy systems: basics, technology and operation. Green Energy Technol. 90, 111–117.

Chapter 8

Hydropower Generation and Its Related Impacts on Aquatic Life (Fisheries)

Berchie Asiedu*, Francis E.K. Nunoo[†], Elliot H. Alhassan[‡], Patrick K. Ofori-Danson[†]

*University of Energy and Natural Resources, Sunyani, Ghana, [†]University of Ghana, Accra, Ghana, [‡]University for Development Studies, Tamale, Ghana,

Chapter Outline

1 Introduction   109
2 Importance of Fisheries to the Ghanaian Economy   110
  2.1 Content of the Paper   110
3 Materials and Methods   110
  3.1 Study Area   110
  3.2 Document Analysis   113
4 Analysis of Impacts of Hydropower Development on Fish and Fisheries   113
  4.1 Impacts of Impoundments on Riverine Ecosystems   113
4.2 Anticipated Impacts of the Bui Dam on Biodiversity   114
4.3 Food Security and Fish Consumption   115
4.4 Fish Diversity   116
4.5 Fish Migration   116
4.6 Fishers' Migration   116
4.7 Fishing Effort   116
4.8 Other Social and Economic Impacts   117
5 Conclusion   117
References   117

1 INTRODUCTION

Dams interrupt streamflow and generate hydrological changes along the integrated continuum of river ecosystems (Vannote et al., 1980; Junk et al., 1989) that ultimately can be reflected in their associated fisheries. Dams may negatively impact on the overall fisheries' productivity (Holcik and Bastl, 1977; Welcomme, 1976, 1985, 1986; Junk et al., 1989). The most obvious effects from placing dams on rivers result from formation of new lentic or semilentic environments upstream from the dam, and tailwater environments downstream from the dam.

Sustainable Hydropower in West Africa. https://doi.org/10.1016/B978-0-12-813016-2.00008-3

A number of water bodies that have been dammed have various potential fish yields (Table 1).

On Lake Volta, tilapias are a major component of the harvest, with catches influenced by water level (higher catches when water level is low).

During reservoir drawdowns, standing timber is harvested for firewood and to facilitate beach seining activities. However, standing timber in the reservoir basin is important for periphyton production. Braimah (1995) estimated that 52% of the fish caught were dependent on invertebrates exploiting this periphyton. Removal of standing timber, in conjunction with overfishing, has negatively impacted the fish stocks.

2 IMPORTANCE OF FISHERIES TO THE GHANAIAN ECONOMY

The fishery sector in Ghana principally encompasses marine fishery, inland (fresh water) fishery, and aquaculture fishery as well as related activities in fish storage, preservation, marketing, and distribution. Fisheries constitute an important sector in national economic development. The sector is estimated to contribute 1.5% of the total GDP and 5% of the GDP in agriculture. About 10% of the country's population is engaged in various aspects of the fishing industry.

Marine fisheries account for over 80% of the fish consumed in Ghana. However, recently, fresh water fisheries including aquaculture is increasingly contributing considerable share of the supply and consumption trends. The structure of the marine fishing industry in Ghana is described by the activities of four identifiable groups within the industry, namely, the Artisanal, Semiindustrial (inshore sector), Industrial (deep sea), and Tuna fleets.

Aquaculture has only recently been adopted as an assured way of meeting the deficit in Ghana's fish requirements. The aquaculture subsector comprises largely small-scale subsistence farmers who practice extensive aquaculture in earthen ponds in contrast to the intensive practices of commercial farmers.

2.1 Content of the Paper

Following the introduction section are materials and methods. This is followed by analysis of the impacts of dams on fisheries resources and what can be done so that livelihoods of thousands of people depending on fisheries are not deprived.

3 MATERIALS AND METHODS

3.1 Study Area

Fig. 1 shows the Bui dam area, which is the major focus of this study. However, other areas with dam construction are also taken into consideration.

TABLE 1 Comparison of the Morphoedaphic Index (MEI) and Estimated Potential Fish Yields of Bui Reservoir With Those of Some African Lakes

Reservoir/Lake	Geographical Region	Estimated or Expected Surface Area (km^2)	MEI	Estimated Potential Fish Yield (kg ha^{-1} year^{-1})
Bui (Ghana)	West Africa	444	0.24	6.30
Volta (Ghana)	West Africa	8700	4.51	15.55
Volta (Ghana)	West Africa	8700	6.1	32.77
Nasser-Nubia (Egypt)	North Africa	6850	9.2	40.40
Kainji (Nigeria)	West Africa	1270	6.6	34.6
Kariba (Zimbabwe)	South Africa	5580	2.8	23.2

Based on Ofori-Danson, P.K., Kwarfo-Apegyah, K., Atsu, D.K., Berchie, A. and Alhassan, E.H., 2012. Final Report on stock assessment study and fisheries management plan for the Bui reservoir. A report prepared for the Bui Power Authority, Accra, Ghana, 106pp.

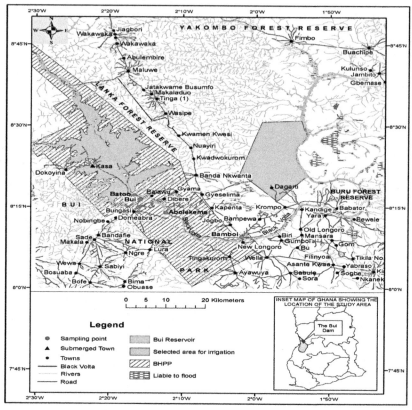

FIG. 1   The Bui dam area.

The idea to develop the Bui dam on the Black Volta at the Bui gorge was conceived in the colonial days of the 1920s when one Albert Kitson, a British-Australian geologist and naturalist on his assignment with the Geological Survey Department of Ghana, visited the site (Ampratwum-Mensah, 2011). The development of a hydropower scheme on the Black Volta at the Bui gorge had been the subject of many studies, namely, detailed studies by J.S. Zhuk Hydroproject of the USSR in 1966, a feasibility study by Snowy Mountains Eng. Corp (SMEC) of Australia in 1976, and another feasibility study by Coyne et Bellier of France in 1995. The 400-MW Bui hydropower scheme was considered to be the most technically and economically attractive hydropower site in Ghana after the Akosombo and the Kpong hydropower plants. The feasibility study of 1995 was subsequently updated by Coyne et Bellier in October 2006 to enable commencement of the project. The Bui Hydroelectric Project was designed primarily for hydropower generation. It however in cludes the development of an irrigation scheme for agricultural development and presents an opportunity for enhanced ecotourism and fisheries. It also includes a Resettlement and Community Support Program. The project was finally completed in 2013 (BPA, 2016).

3.2   Document Analysis

A number of publications and reports produced by individuals, organizations, and projects were analyzed as secondary data sources (i.e., Bui Power Authority, Volta River Authority, Ministry of Fisheries and Aquaculture Development, Food and Agricultural Organization). Data were gathered on dam statistics, fish stocks, hydrology, and climatic conditions.

4   ANALYSIS OF IMPACTS OF HYDROPOWER DEVELOPMENT ON FISH AND FISHERIES

4.1   Impacts of Impoundments on Riverine Ecosystems

Building a dam across a river, and impounding water behind it, may cause profound changes in the limnological regime and biological productivity of the water body (Egborge, 1979; Reynolds, 1997; Ogbeibu and Oribhabor, 2002). The ecological impacts of impounding a river have been dramatic and extensive. Dams can affect the geomorphology of streams that have a large sediment load, as the reservoir traps sediments and releases clear water. The resulting downstream geomorphic effects of clear water releases from dams include channel instability and alteration of habitat (Collier et al., 1996). Dams can also affect fish by blocking upstream and downstream passage. Elimination or reduction of spawning grounds, or delayed access to the spawning areas has been the most significant effects of physical barriers (FAO, 2001). The blockage of upstream fish movements by dams may have serious impacts on species whose life history includes migrations for various purposes. For fish trying to move upstream, a dam can pose an impassable barrier, and fish moving downstream are at high risk of being entrained in the turbine intake and injured or killed during downstream passage (FAO, 2001).

Flow variability controls physicochemical and hydrobiological phenomena in a river. Impoundments, particularly those of storage-release nature, reduce the natural variability of flow, although hydropower impoundments may increase diel variability. Some hydropower dams have underground power stations, resulting in the desiccation of a section of the river downstream of the dam. Resident fishes experiencing flow alterations may be affected for great distances downstream. Flow modifications affect water quality, water depth and velocity, substrate composition, food production and transport, stimuli for migration and spawning, survival of eggs, and eventually fish species composition (Petts, 1984).

The physicochemical and hydrobiological attributes of the downstream ecosystem are dictated by whether the release of water is drawn from hypolimnion, epilimnion, or multilevels (Cassidy, 1989). Depth of withdrawals affects water temperature, dissolved gases, nutrients, turbidity, biotic assemblage, and diversity. Hypolimnetic releases are relatively cold, oxygen depleted, and nutrient rich. Epilemnetic releases on the other hand are typically less disruptive as temperature and water quality characteristics are more suitable to the downstream biota (Crisp, 1977). Reducing water flow therefore changes the biota and landscape downstream (Simons, 1979).

Reductions in sediment load caused by impoundments prompt the river downstream to recapture its load by eroding the downstream channel and banks. Release patterns also affect downstream biota in several ways (Welcomme, 1985). Large flow variations may adversely affect downstream productivity by impacting spawning and disrupting benthic populations. Lower nutrient concentrations in releases can result in lower primary production in the tail water. Conversely, nutrient-rich releases stimulate increases and lavish the development of algae and macrophytes. The benthic community may shift toward grazers and collectors and experience loss of diversity as organisms depending on thermal cues for spawning, hatching, and emergence will dwindle (FAO, 2001).

The main effect of impoundment on fish is a shift in species composition and abundance, with extreme proliferation of some species and reduction, or even elimination of others (Agostinho et al., 1999). According to FAO (2001), the level of impact on the biological diversity is generally influenced by the characteristics of the local biota (e.g., reproductive strategies and migratory patterns), characteristics of the reservoir (e.g., morphology and hydrology), design and operational characteristics of the dam, and characteristics and uses of the watershed (e.g., agriculture, mining, and urbanization). Wetzel (1990) noted that the response of fish assemblages to impoundment is a chaotic succession of reactions marked by a reduction in the established interdependence among species, and a lower biotic stability as well as natural succession processes.

Changes in habitat caused by impoundments often limit the lotic fish fauna to the upper, unimpounded reaches of streams. This is because the reservoir acts as a barrier for dispersal, preventing upstream or downstream passage, these populations often remained isolated. These small and fragmented populations may survive for many years in a river basin, but much of the original genetic variation may be lost (Wilson, 1988). Lack of passage also restricts the ability of fish to decolonize suitable habitat following catastrophic events. Thus, impoundments have fragmented the home ranges of certain species, causing local extinctions.

4.2 Anticipated Impacts of the Bui Dam on Biodiversity

The biodiversity of the Black Volta includes the living aquatic resources such as fishes (both shell and fin fishes) and some isolated populations of primates and other large mammals such as the hippopotamus (*Hippopotamus amphibious*) population and plant vegetation that depend on the river at Bui. The major anticipated environmental impact that is expected to affect the flora and fauna in the basin would be the effect of lacusterization on the Black Volta, as a result of the construction of the Bui dam in June 2011.

Construction of the Bui dam and associated structures, and the creation of the reservoir, will cause both loss and alteration of habitats, with resulting impacts on ecology and biodiversity. Filling and operation of the Bui dam will ultimately create $440 \, \text{km}^2$ of new lacustrine habitat with a maximum depth of $29 \, \text{m}$, replacing approximately $40 \, \text{km}$ of riverine habitat along the Black Volta (Ampratwum-Mensah, 2011). Prior to filing, the main part of the reach will retain its current riverine characteristics, with the exception of the temporary diversion channel. During this period, the biological communities in the reservoir will begin to be exposed to lacustrine characteristics. The initial change will include declines in mayflies and certain fish species that prefer running water and coarse substrate. Species that prefer shallow habitat are likely to colonize the periphery of the reservoir, and others that require running water may disappear or persist as relict populations in the headwaters of the reservoir (Environmental Resources Management (ERM), 2007).

Sediment retention and subsequent deposition within the reservoir will cause most of the coarse substrate, rocky outcrop, and other elements of the riverbed to progressively disappear underlayers of silt transported from upstream. This will alter the fish and macroinvertebrate composition with a reduction in the abundance of riverine fishes such as Mormyridae species and an increase in benthic and algal feeders, such as species of Bagridae and Cichlidae (Gordon et al., 2003).

4.3 Food Security and Fish Consumption

The creation of Bui dam has resulted in low consumption of fish by some communities near the reservoir. Fisherfolks in the Bator community near the dam do not prefer the consumption of tilapias for food. This will result in culturing of certain species such as the African catfish and Nile tilapia. Presently, net profit from aquaculture in Ghana is less than 5%. Fishers relocating from their communities as result of creation of dams mean that livelihood and low-protein sources are lost; building of the Bui dam has forced the relocation of about 1300 people, of which about 80% are fishers. As a result of the formation of the Volta Lake, about 78,000 people were relocated to new towns and villages, along with 200,000 animals belonging to them, while about 120 buildings were destroyed.

Livelihoods from fishing have become unreliable. In the Lucene and Agbegikuro communities around the Bui dam area, there have been an increase of about 20%–30% in fishing activities after the construction of the Bui dam (Atindana et al., 2015). However, this is not stable throughout the year and decreases drastically during the lean season (November–March).

4.4 Fish Diversity

According to Ofori-Danson et al. (2012) before the impoundment of the Black Volta by the Bui dam, five species formed major components of the fish landings. The species were *Chrysichthys auratus (family Claroteidae), Alestes baremoze, Hydrocynus forskalii (family Alestedae), Hemisysnodontis membrananceus (family Mochokidae), Labeo coubie (family Cyprinidae)*, and *Lates niloticus (family Centropomidae)*. These species have not differed after impoundment was over. The effects of water resource development on freshwater biodiversity have been widespread in the United States and Canada, where 40% of the estimated 750 freshwater fishes and 72% of the 297 mussels are extinct, endangered, or threatened (Kottelat and Whitten, 1996).

4.5 Fish Migration

Fish populations are highly dependent upon the characteristics of the aquatic habitat, which supports all their biological functions (Marmulla, 2001). The construction of dams can block or delay upstream fish migration and thus contribute to the decline and even the extinction of species that depend on longitudinal movements along the stream continuum during certain phases of their life cycle. Mortality resulting from downstream passage through hydraulic turbines or over spillways can be significant.

Changes in discharge regime or water quality can also have indirect effects upon fish species. Increased upstream and downstream predation on migratory fish is also linked to dams, fish being delayed and concentrated due to the presence of the dam and the habitat becoming more favorable to certain predatory species (Marmulla, 2001).

4.6 Fishers' Migration

The creation of dams results in people migrating to the reservoir areas. There has been unexpected increase in human population in the Bui dam area. This could lead to additional fishermen migrating from below the dam. If this migration is not controlled, the fisheries of the Bui reservoir could experience explosive overexploitation of the fisheries and hence erode fisheries productivity benefits and conservation efforts within a short period. The increment in population is resulting in pressure on existing few amenities, such as hospitals, schools, and roads as well as other social vices (e.g., prostitution, armed robbery, theft).

4.7 Fishing Effort

Fishermen are complaining about the increased effort needed to obtain the same fishing yield as 10 years ago before the river was impounded. Fishermen are of the opinion that the average size of fish has become considerably smaller in the last few years. A community-based reservoir fisheries management program can offer a solution.

4.8 Other Social and Economic Impacts

World Bank estimates that globally 10 million people are displaced each year due to dam construction, urban development, and transportation and infrastructure programs (World Bank, 1995). This number is shockingly high, but it still fails to account for large numbers of the displaced. Displacement tallies almost always refer only to persons physically ousted from legally acquired land in order to make way for the planned project, ignoring those living in the vicinity, or downstream from, projects, whose livelihoods and sociocultural milieu might be adversely affected by the project (Scudder, 1996).

However, there are also many documented cases of dam operation adversely affecting the livelihoods and health of people, living not just in the immediate vicinity of the dam, but sometimes many hundreds of kilometers downstream (McCartney, 2007).

Resettlement issues can be classified into two categories, namely, voluntary and involuntary. Involuntary resettlement has been a companion of major development projects or programs throughout history, and has been permanently written into the evolution of industrial as well as developing countries (World Bank, 1995).

Over the years, roughly 80 million to 90 million people have been relocated as a result of infrastructure programs for dam construction, sanitation infrastructure, urban upgrading, and transport improvement (World Bank, 1995). Their livelihoods are lost as a result of these constructions. Ultimately, there is loss of culture and social integration as a result of people relocating from the affected areas.

5 CONCLUSION

Hydropower development generation has negative impact on fish and fisheries. As a result of dam construction, there are changes in the ecosystems, fish diversity, and species landed. The fishing effort and migration also increase as a result of dam constructions. Thousands of livelihoods are lost due to dam construction. It is recommended that environmental and social impact assessment be carried out prior to dam construction. Additionally, fisheries advisory board should be established in areas where dams are constructed so that optimum economic benefits can be derived from the fishery resource.

REFERENCES

Agostinho, A.A., Miranda, L.E., Bini, L.M., Gomes, L.C., Thomas, S.M., Susuki, H.I., 1999. Patterns of colonization in neotropical reservoirs and progress on ging. In: Tundisi, J.G., Laiden, M. (Eds.), Theoretical Reservoirs Ecology and It Application. Backhuys Publishers, The Netherlands.

Ampratwum-Mensah, A., 2011. The Bui hydroelectric power project-the bare facts. Daily Graphic 3, 21.

Atindana, S.A., Mensah, P., Alhassan, E.H., Ampofo-Yeboah, A., Abobi, S.M., kongyuure, D.N., Abarike, E.D., 2015. The socio-economic impact of Bui dam on resettled communities; a case study of Lucene and Agbegikuro communities in the northern region of Ghana. UDS Int. J. Dev. 2 (1), 1–51.

Braimah, L.I., 1995. Recent developments in the fisheries of Volta Lake (Ghana). In: Crul, R.C.M., Roest, F.C. (Eds.), Current Status of Fisheries and Fish Stocks of the Four Largest African Reservoirs: Kainji, Kariba, Nasser/Nubia and Volta. FAO, Rome. CIFA Technical Paper No. 30.

Bui Power Authority (BPA), 2016. Project Background. http://www.buipower.com/node/9. (Cited 10 December 2016).

Cassidy, R.A., 1989. Water temperature, dissolved oxygen and turbidity control in reservoir releases. In: Gore, J.A., Petts, G.E. (Eds.), Alternatives in Regulated River Management. CRC Press, Boca Raton, FL.

Collier, M., Webb, R.H., Schmidt, J.C., 1996. Dams and Rivers: A Primer on the Downstream Effects of Dams. In: United States Geological Survey Circular 1126. Tucson, AZ, 94 pp.

Crisp, D.T., 1977. Some physical and chemical effects of the cow green (upper Teesdale) impoundment. Freshwater Biol. 7, 109–120.

Egborge, A.B.M., 1979. The effect of impoundment on the water chemistry of lake Asejire, Niger. Freshwater Biol. 9, 403–412.

Environmental Resources Management (ERM), 2007. Environmental and Social Impact Assessment Study of the Bui Hydroelectric Power Project. 204 pp. http://www.erm.com (Retrieved 12 March 2013).

FAO (2001). Dams, Fish and Fisheries, Opportunities, Challenges and Conflict Resolution. Marmulla, G.(ed). FAO Fisheries Tech. Pap. No. 419. Rome, Italy, 166 pp.

Gordon, C., Ametekpor, J.K., Koranteng, K., Annang, T., 2003. Aquatic ecology component. In: Bui Hydroelectric Power Project. FAO Fisheries Technical Paper No. 419.

Holcik, J., Bastl, I., 1977. Predicting fish yield in the Czechoslovakian section of the Danube River based on the hydrological regime. Int. Rev. Gesamten Hydrobiol. 62 (4), 523–532.

Junk, W.J., Bayley, P.B., Sparks, R.E., 1989. In: Dodge, D.P. (Ed.), The flood pulse concept in river-floodplain systems. Proceedings of the International Large River Symposium. 106. Canadian Special Publication of Fisheries and Aquatic Sciences, Ottawa, pp. 110–127.

Kottelat, M., Whitten, T., 1996. Freshwater Biodiversity in Asia: With Special Reference to Fish. World Bank Technical Paper, 343.

Marmulla, G., 2001. Dams, fish and fisheries. Opportunities, challenges and conflict resolution. FAO Fisheries Technical Paper. No. 419. FAO, Rome, 166 pp.

McCartney, M., 2007. Decision Support Systems for Large Dam Planning and Operation in Africa. International Water Management Institute, Colombo.

Ofori-Danson, P.K., Kwarfo-Apegyah, K., Atsu, D.K., Berchie, A., Alhassan, E.H., 2012. Final Report on Stock Assessment Study and Fisheries Management Plan for the Bui Reservoir. A report prepared for the Bui Power Authority, Accra, Ghana, 106 pp.

Ogbeibu, A.E., Oribhabor, B.J., 2002. Ecological impact of river impoundment using benthic macro-invertebrates as indicators. Water Res. 36, 2427–2436.

Petts, G.E., 1984. Impounded Rivers. John Willey and Sons, Chichester. 326 pp.

Reynolds, C.S., 1997. Vegetation process in the pelagic: A model for ecosystem theory. Inter Research Science Publishers, Oldendorf, Luke, Germany. 403 pp.

Scudder, T., 1996. Development-Induced Impoverishment, Resistance and River-Basin Development. In: McDowell, C. (Ed.), Understanding Impoverishment. Berghahn Books, Providence, Oxford.

Simons, D.B., 1979. Effects of Stream Regulation on Channel Morphology. In: Ward, J.V., Stanfurd, J.A. (Eds.), The Ecology of Regulated Streams. Plenum Press, New York, NY.

Vannote, R.L., Minshall, G.M., Cummins, K.W., Sedell, J.R., Cushing, C.E., 1980. The river continuum concept. Can. J. Fish. Aquat. Sci. 37, 130–137.

Welcomme, R.L., 1976. Some general and theoretical considerations on the fish yield of African rivers. J. Fish Biol. 8, 351–364.

Welcomme, R.L., 1985. River Fisheries. FAO Fish. Tech. Pap. No. 262. FAO, Rome. 330 p.

Welcomme, R.L., 1986. The effects of the Sahelian drought on the fishery of the Central Delta of the Niger River. Aquacult. Fish. Manag. 17, 147–154.

Wetzel, R.G., 1990. Land-water interfaces: metabolic and limnological regulators. Verh. Internat. Verein. Limnol. 24 (6), 24.

Wilson, E.O., 1988. The Current State of Biological Diversity. National Academy Press, Washington, DC.

World Bank, 1995. Resettlement and Development: The Bank Wide Review of Projects Involving Involuntary Resettlement, 1986–1993. Environment Department Paper No. 032, Resettlement Series, Washington, DC.

Chapter 9

Socioeconomic Impacts of the Bui Hydropower Dam on the Livelihood of Women and Children

Nina Schafer, Heidi Megerle, Amos T. Kabo-bah
Hochschule für Forstwirtschaft, Rottenburg, Germany

Chapter Outline

1. **Introduction**	**121**	3.3 Waste Management	131
1.1 Bui Hydropower Dam	122	3.4 Lost Tourism in the Bui	
2. **Methodology**	**124**	National Park	133
3. **Results**	**125**	4. **Discussion**	**133**
3.1 Changes in Livelihood	127	5. **Conclusion**	**135**
3.2 Water Situation	129	**References**	**136**

1. INTRODUCTION

"Over 45,000 times in the last century, people took the decision to build a dam. Dams were built to provide water for irrigated agriculture, domestic or industrial use, to generate hydropower or help control floods" (WCD, 2000, p. 1). Every year $3800 \, \text{km}^3$ of fresh water is taken from the world's lakes, rivers, and aquifers (WCD, 2000, p. 3). Growing population, economic activities, and nonsustainable irrigation systems increase the human demand for water and energy. Especially in developing countries, energy supply is essential for economic growth, stability, and human welfare also concerning attaining the MDGs (Millennium Development Goals) (UNEP, 2010, p. 147). A better access to energy will improve food security through refrigeration, as well as education development through electric lighting and will reduce the use of wood for fuel (UNEP, 2010, p. 148). However, before building a dam, the benefits and disadvantages must be weighed carefully and the project needs to get planned well. According to the World Commission on Dams (WCD, 2000, p. 2), "[…] the end of any dam project must be the sustainable improvement of human

Sustainable Hydropower in West Africa. https://doi.org/10.1016/B978-0-12-813016-2.00009-5

welfare." That means the dam should help to progress living condition, sustainable economy, and environment as well as social equality.

Unfortunately, most of the time, negative impacts on an affected area's economy, lifestyle, and environment overshadow the benefits of large dams. Hydro-dependent countries, most of them located in Africa, are suffering because of missing rainfalls and drought events in recent years. Blackouts are the result and many dams fail to meet their capacity targets, which in turn has a negative influence on the economic situation of the country (Hildyard, 2008). Moreover, environmental consequences of large dams are numerous and varied and include direct impacts to the biological, chemical, and physical properties of rivers and streamside environments. According to the WCD (2000, p. 92), large dams led to a loss of forest and wildlife habitats, as well as a loss of species populations and agricultural fields at the upstream catchment area. Consequently, rural fishers, who depend on fisheries either as a primary or a supplementary source of living, lose their source of income, likewise farmers losing their farmlands. Also, reported cases of water-related diseases such as schistosomiasis and malaria close to the reservoir are not unusual impacts of large dams especially in tropical areas. To create these reservoirs, huge areas of land have to be flooded, where people are living and cultivating the land. These people need to be resettled and are entitled to receive full compensations. Unfortunately, these compensations do not always reach everybody affected by the project or do not fulfill all the promises authorities made (UCS, 2013). All in all, giving just a few points of negative impacts of dam constructions shows that socioeconomic issues got less attention when dams were built.

1.1   Bui Hydropower Dam

The impacts of large dams described are not different from those at the Bui Hydropower Dam on the Black Volta Basin in Ghana. The dam generates 400 MW and is located at the Bui Gorge at the southern end of Bui National Park in Ghana. The Bui Hydroelectric Power Project is approximately 150 km upstream of Lake Volta and its aim should be to increase the electricity generation capacity in Ghana by 22%, up from 1920 MW in 2008 to 2360 MW (Tornyie, 2015, p. 2). The specific location of the Bui Project is particularly suitable because of the deep gorge where the Black Volta River flows through the Banda Hills. The project includes a main dam in Bui Gorge and two smaller saddle dams in the neighboring Banda Hills, which creates the reservoir. Saddle dams are constructed between topographic low points along the reservoir lines to prevent the reservoir from spilling out of the basin. The dam impounds a reservoir with a full supply level of 183 m and a surface area of 440 km². The new transmission lines will run east from the project to Kintampo and south to Techiman and toward Bolgatanga, Sunyani, and Wa (ERM, 2007, p. 4–6) (Fig. 1).

The dam had been planned since 1920 and was part of a bigger dream imagined by Ghana's first president Kwame Nkrumah to industrialize and

FIG. 1   Location of the Bui Dam (http://www.planwallpaper.com/static/images/ghana-and-ghana-in-africa.jpg 2014; BPA, 2015).

modernize Ghana and Africa as a whole. The project was revived in 2002 under the leadership of the former President of Ghana, John Kufuor. In the year 2007, John Kufuor established an Authority, the Bui Power Authority (BPA), to plan, manage, and execute the Bui Dam project. The first task, the BPA had, was to obtain the "Environmental Permit" of the Environmental Protection Agency (Hensengerth, 2011, p. 14). The Government of Ghana commissioned the consultancy firm Environmental Resources Management (ERM) to find out about the environmental and social impacts caused by the construction and operation of the Bui project. This was done to avoid negative impacts on affected people and moreover to improve their livelihood. An Environmental and Social Impact Assessment (ESIA) Report, the Resettlement Planning Framework (RPF) and an Environmental and Social Management Plan (ESMP) resulted out of this study (Raschid et al., 2008, p. 10). These plans provided the basis on which the Ghanaian EPA was considering its permitting of the Bui Project (ERM, 2007, p. 1).

The main construction started in 2009 and promised to develop Ghana's electricity supply, especially in some parts of the northern regions. The project has been a collaboration between the government of Ghana and Sino Hydro, a Chinese construction company. The total project costs are estimated to be around US\$622 million. A small part of the costs is financed by Ghana's own resources (about US\$60 million). The rest is being financed by two credits of the China Exim Bank. "The proceeds of 30,000 tons per year of Ghanaian cocoa exports to China, which are placed in an escrow account at the Exim Bank, serve as collateral for the loan" (Tornyie, 2015, p. 2). In December 2013, the dam and the power station were inaugurated by President John Mahama. Since then, 85% of the profits of electricity sales from the hydropower plant have been going straight to the escrow account (Tornyie, 2015, p. 3).

"There were other alternatives to solving the power crises by concentrating on other major dams and managing power to avoid power wastage but the government chose to build a new dam" (Tornyie, 2015, p. 1). About 1200 people had to leave their homes and more than 20% of the Bui National Park was flooded, just to name the main consequences of building the dam. Less attention has been given to the local people's participation, in order to influence the decisions made. They have been relegated to the background instead of being key participants in an issue that is about their lives (Tornyie, 2015, p. 3). None of their livelihood activities—farming, fishing, trading, and tourism, and their assets of fertile farmlands and fishing grounds—have been completely restored till now. Also, the access to their source of water, the Black Volta, both for drinking water and domestic use is affected and must be compensated for.

The question is how life changed for the affected people after the dam construction, and whether promised compensations have improved their lifestyle. Several types of studies of the issue of livelihood assessment impacts of the Bui Dam development across the study area have been carried out by interviewing the leaders of households—men. However, studies with vulnerable groups like women and children are rare, although they are the ones suffering most as a result of the loss of livelihood and raising unemployment. Therefore, one hundred questionnaires were given especially to women in four different communities in the Bui Dam area. Special attention was given to the changes in their livelihood, water supply, and sanitation systems before, during, and after the dam was constructed.

2. METHODOLOGY

Quantitative and qualitative methods, as well as a literature review, were used to obtain all the relevant data and determine the socioeconomic impacts of the Bui Dam on women and children. For this purpose, questionnaire-based interviews, observations, and one in-depth interview were conducted in the Bui Dam area, which is located on the borders of the Northern and the Brong-Ahafo Regions in Ghana.

Therefore, books, publications, theses, reports, and Bui Dam-related plans of the BPA and the Government of Ghana were read and analyzed first. This was done to assess which sources could be of importance for the research and to decide whether or not to include them in the research. Of particular importance are the RPF and ESMP by the US consulting firm Environmental Resources Management for a comparison with the collected data, as well as the WCD guidelines to give a good overview of large dams in general.

The data for the study were gathered from two main sources, namely, primary sources and secondary sources. The primary data for the research were obtained through both quantitative and qualitative sources. Quantitative sources were used to generate data of a larger sample population that could be transformed into useable statistics (Wyse, 2011). In this case, the quantitative data collection method was questionnaires. Stakeholders from whom the questionnaires were obtained included women and men, living within the study communities in the Bui Dam area. Qualitative data were used to gain a better understanding of the impacts of the resettlement and dive deeper into the problem. Observations and an individual interview with the Assembly man of Resettlement Part B Maxwell Gbadego represent the qualitative data methods used for this report.

The data collection included issues of the affected people in four different communities and their perceptions about water quality, access to water, sanitation facilities, waste management, and changes of livelihood. People were asked about these issues, and in particular to describe the situation before, during, and after the dam was constructed. This helped to find out, how life really changed, especially with regard to women and children in the communities. 73 out of the 100 completed questionnaires were answered by women affected by the dam. However, in each community, men were questioned as well. Men, as the heads of the households, had more knowledge about the resettlement actions, and could give more detailed information about the livelihood changes for fishermen and farmers.

The collected data from the questionnaire have been analyzed with the Statistical Package for Social Science (SPSS) and MS Excel. Descriptive statistics, such as frequencies and crosstabulations, were used to display and analyze the results. Information generated through the in-depth interview and own observations were described and added to the respective chapter.

3. RESULTS

Both the Akosombo and the Kpong Dam resettlement schemes failed in the past; it was therefore really important for the Bui Dam resettlement to be successful and sustainable in order to restore the affected inhabitants´ livelihoods and in the best case even improve their situation. To achieve this goal, the Resettlement Planning Framework (RPF) was developed by the Bui Power Authority to ensure a successful resettlement of the affected communities and create economic opportunities for the affected people of the project (Miine, 2014, p. 1).

The Bui Project forced eight villages with a total population of 1216 affected people by the construction of the dam and the reservoir to resettle. New resettlement townships were given to the families as well as communal facilities, like community centers, nursery school, a place of worship, boreholes and Kumasi Ventilated-Improved Pits (KVIPs). A compensation payment of GHS 100 was paid to each member of the affected communities and GHS 50 to plough new farms, which was never enough, considering the fact that new tree crops need around 6 years until they bear fruits again. The BPA further acquired two acres of land for each household resettled under Part A (see later). Moreover, the affected people got an income support of GHS 100 per month for each household for the first year, which is a too short time frame in order to become self-sufficient (BPA, 2015) (Fig. 2).

The implementation of the resettlement program has been divided into three parts.

Phase A: Jama, the host community, is located close to the construction of the Bui Generating Station. Its population is about 1500 people, whose

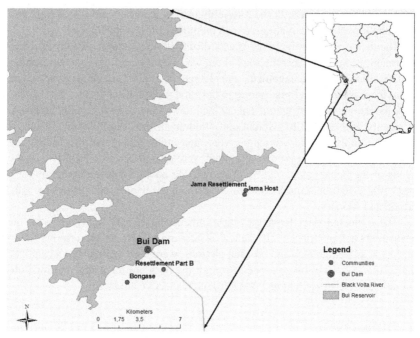

FIG. 2 Selected villages in the study area.

livelihood activities are farming, hunting, fishing, and trading. The Resettlement Township, known as Jama Resettlement, has a population of 217 people within close proximity to the Bui Generation Station. Villages close to the dam site had to be resettled first, because this is where the construction activities started first. The Jama Resettlement Township was equipped with a community center, solar-lighting systems, an access to the road, two pump boreholes, toilet facilities, a nursery school, and 50 housing units (BPA, 2015).

Phase B involved the relocation of three communities that originally settled within the reservoir area. The three affected villages have a total population of 899 people, who were relocated to a new township in 2011. The Resettlement B township is located between Bongase and the Bui Generating Station. The affected young people of these villages have been assisted by the BPA to work as laborers, masons, and carpenters during the construction of their new homes to earn some extra money.

Phase C covers the personnel of the Game and Wildlife Division living at the old Bui Camp (BPA, 2015).

However, also a lot of people who have not been resettled are still affected by losing their farmland, such as Bongase, by the noise of transmission lines and changes of the river flow downstream the dam (Hensengerth, 2011, p. 25).

An overview of the number of interviews in the respective villages is shown in Fig. 3.

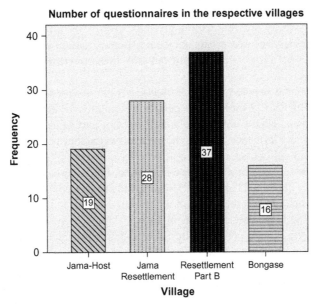

FIG. 3 Number of questionnaires in the respective villages.

3.1   Changes in Livelihood

Before drawing conclusions about how the life changed for the affected people, it is important to look at their livelihood before the dam was constructed. According to the respondents, most of them are living from subsistence farming and were trading the crop yield on the local markets. The answers of the people from Resettlement Part B varied a bit. Most of the men were fishers and their wives were selling smoked fish, explained by the fact that their previous villages were really close to the Black Volta. During the construction, there were no significant changes in the activities of the affected people. The only exception was that few women took the chance to sell food for the workers, who were working at the construction site; this was one of the few positive impacts of the Bui Dam for the women.

The current situation is alarming and the first negative impacts of the Bui Dam can be seen. Only a few women are still going to the markets to sell their goods, because what they harvest from the farms is just enough to feed their families with. Furthermore, the unemployment rate rose rapidly. Women and especially men in Resettlement Part B lost their jobs, as fishing and trading fish was their main source of livelihood. According to the fishermen, the BPA does not allow them to fish everywhere and the few places they are allowed to fish in, are far away from their homes and not abundant with fish. Maxwell, the Assembly man of Bongase and Resettlement Part B, reported on another phenomenon, which affects especially the fishermen. An estimated number of 30,000 people from Mali, Burkina Faso, and Cote d'Ivoire are settled along the

Black Volta and the Bui Dam in illegal housings. These people are tolerated, but they take away possible fishing grounds, destroy the environment with illegal gold mines, and exacerbate the local fishers' situation as well. During the research, it was not possible to find out why these migrants are tolerated by the BPA.

Asking if the people had any influence on the resettlement process, 57% answered with "No." Especially the people in Bongase who have lost their farmlands, and Jama Host Community, which was not part of the relocation itself, did not have any influence at all. Some even answered they heard about it on the news or radio. Of course, these people who have lost their villages and moved to new places got to know about the relocation, but were faced with a fait accompli. They explained that the BPA was coming to their villages and told them about the Bui Dam and the reservoir and promised them to improve their standards of living in the new resettlement areas.

On the other hand, people really like the new infrastructure and their new homes. The roads are in a much better shape and the few public buildings, like schools, nurseries and, health centers, which have already been set up in the communities, are much appreciated. Also, a more trained staff is available, such as teachers and nurses. Furthermore, having access to electricity improves theoretically the living conditions of the families, even though most of them do not have enough money to pay their bills (Fig. 4).

However, many people complained that most of the promised compensations, such as schools, nursery, and more health care centers, are still missing or that the number of facilities is not enough. Especially compensations on new farmlands have not been sufficient. The new farmland is located far away and going there every day for cultivating the soil and planting is too much work to do. "We are tired. The distances are too long, so we cannot go there every day," a resident of Bongase explained. Due to the fact that farmers are not close to their farms anymore and are not going there every day, wild animals destroy their farms. They also mentioned that the rainfall really changed, which affects the infertile land even more. A resident of Resettlement Part B complained that "no rain reaches our communities since the forest is gone, who caught the rain.

FIG. 4 Old resettlement (BPA, 2015) and new resettlement (author's own picture, 2016).

To date, companies still cut down the big trees in the park to sell the timber." Moreover, the BPA paid GHS 100 to each household for the first year to support their income. However, most of the people, especially farmers, complained that 1 year is just too short in order to be supported well: "I am a cassava farmer. After the dam was constructed I lost all my land and was provided with a new land. But it takes two years to harvest the cassava. So I had no income for two years. The BPA should have taken care of us for a longer period. One year is not enough, considering the number of people we are. We are seven people in one house and we got GHS 100 per month, which is not even enough to feed my children." Maxwell further mentioned another problem of the compensation. For each cash crop that a farmer lost, they got GHS 8 for a lifetime, excluding that a cashew tree, for example, provides a farmer with GHS 30–40 per annum, which was a safe income for the whole family.

The dam also implicated some other problems. During the survey in Bongase, a few "mixed-race" children attracted my attention. As a reply to my question, people explained, that hundreds of Indians and Pakistanis worked at the dam site during the construction. People reported of cases (in Bongase and Resettlement Part B), in which Ghanaian women had relationships with these workers, who promised to take them to India and Pakistan for a better life, which they never did. They left the women with the children behind and did not take care of them anymore. Cases of prostitution and violations in the communities close to the dam site have been reported and are a well-known impact on women on large projects like this in these areas.

As a result of all these changes and negative impacts on their livelihood, about 31% of the respondents answered that they know of cases where people had left the communities already. The reasons for this are unemployment, losing fertile farmlands, not enough fishes, and fear of the future. Those are reasons, which should not have taken place, according to the BPA and the RPF.

3.2 Water Situation

One of the main goals of the BPA was to improve the water situation for the affected people. Before the dam was constructed, 67% of the respondents were fetching their water from the Black Volta, since they lived next to it, with the exception of Jama Host Community and Bongase. Jama Host Community and Bongase have already used boreholes to secure their water demand. Bongase further fetched water from other streams due to the distance from the Black Volta. During the construction, right through the completion of the Bui Dam, fetching water from the Black Volta was replaced by boreholes. In all settlements, boreholes became the only source of improved drinking water. Everybody has now a shorter distance to go to reach a borehole in their community. Most of them are going there in the morning and evening; others go there, whenever they need water during the day. The only difference to earlier times in the previous villages is the duration of fetching water. Whereas it used to take some minutes to get water from the river before the dam was constructed, many people reported on waiting times now. The number of boreholes is not enough for the number of people living in the communities, so that especially in the evening people have to wait up to 1 or 2 hours to get their full amount of needed water. Also, asking for the main method of water collection, the answer showed an unambiguous result: 97% of the respondents mentioned that the main method of the collection is the head (see Fig. 5).

However, the question arose how people in the communities feel about the quality of water before, during, and after the dam was constructed. People who used the Black Volta before the Bui Dam was constructed, indicated that the river water of the Black Volta was really clean and tasty, however really muddy in the rainy season. The few people using boreholes even before the dam had been built reported that sometimes the borehole water was a bit salty, but in general really good. In the streams, which were used by the inhabitants of Bongase, the Guinea worm disease, and dirty water was affecting the people. The Guinea-worm disease "is a crippling parasitic disease caused by the Dracunculus medinensis—a long thread-like worm. It is transmitted exclusively when people drink stagnant water contaminated with parasite-infected water fleas" (WHO, 2015).

During the construction, the ones who already used boreholes as their source of water reported that the water had a good quality but smelled sometimes, as a result of the drilling operations at the construction site. The few who still used the Black Volta or other streams to fetch their drinking water from, mentioned that the water was not clean like it used to be and they could not use it any longer for drinking water. Everybody mentioned the black flies that increased a lot at the beginning of the dam construction. The flies have resurfaced at the Bui Dam project area and especially bothered the workers and the residents who could not go out without protecting their bodies and especially their faces to prevent the bites. Black flies, which breed solely in fast flowing waters, can cause River Blindness (Barry et al., 2005, p. 11). "As part of its social corporate responsibilities, the BPA initiated programs to militate against the black flies at

FIG. 5 Women collecting water in Jama Host Community and Jama New Resettlement (author's own picture, 2016).

Bui and its environs are yielding positive results with the total eradication of the black fly nuisance nearly in sight" (Modern Ghana, 2015). The spraying was successful and also the affected people confirmed that the flies reduced during the construction. Asking for the current situation and the water quality, 96% of the respondents answered that they are real pleased with the borehole water, just in some cases it tastes a bit salty. Unfortunately, many people mentioned that the black flies are coming back again especially at places around the boreholes and of course close to the dam. It can only be hoped that the BPA continues to spray to banish these flies.

3.3 Waste Management

Before starting the research in the Bui area, there was the suspicion that probably most of the solid waste from the houses and also waste water from the sanitary facilities would be discharged into the river, since most of the settlements have been close to the river before. The river plays a religious role in some people's life. Moreover, they see the river as their source of life because they are drinking the water and ensuring their livelihood with the fishes caught in the river. This is why they worship the river and do not pollute it.

Before the dam was constructed, most of the villages owned a refuse dump, where people sent their solid waste from the house, while others were using the bush. A correct waste management did not take place thus far. According to the respondents, the same "waste management" has been practiced while the dam was constructed. A significant change in the waste disposal can be seen after the resettlement. While Jama Host Community and Jama Resettlement are still using the refuse dump or bush for their solid waste from the house, some of the families in Bongase and Resettlement Part B started to use big dustbins to collect their waste. Zoomlion Ghana Limited (2016), a Ghanaian company for environmental sanitation services, empties these trash cans in the villages.

Looking at the sanitation facilities the people used before the dam was constructed, it turns out that KVIPs' self-made pit latrines and the bush itself were used to the same extent. The only improved sanitation facility is the KVIP (see Fig. 6), which confirms the bad sanitation situation in Ghana.

But also during the construction, the situation did not really change for the better. More people were using the KVIP due to the fact that the new resettlements were provided with KVIPs as well. Interesting is the current situation. To summarize, 51% of the respondents are using the KVIPs located in the communities, but still 40% are using the bush, even though all communities are provided with improved facilities, for which there may be various reasons.

In Bongase, for example, the BPA did not yet fulfill their promises to provide the inhabitants with public KVIPs. Residents reported that the numbers of KVIPs are not enough and the ones they have in their community were donated by volunteers and not by the BPA. Furthermore, they like going to the bush to ease themselves, even though they need to walk a couple of minutes. One

FIG. 6 Public KVIPs in Jama New Resettlement (author's own picture, 2016).

reason could be that they like to avoid the smell in the KVIPs. In Resettlement Part B, the same situation is found. The public KVIPs are not enough or need to be pumped out and although their houses have all been provided with a toilet, they prefer going to the bush or using their own self-made pit latrine. The same phenomenon was reported in Jama Resettlement. Most of the residents of the Jama Host Community are using public KVIPs (see Fig. 7). The BPA increased the number of toilets when people joined their community during the construction of the dam.

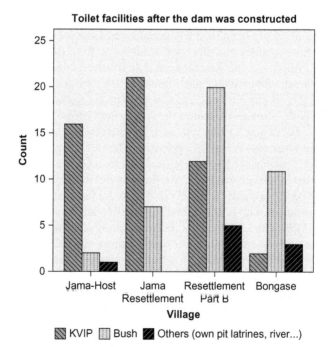

FIG. 7 Overview of the toilet facilities after the dam was constructed.

3.4 Lost Tourism in the Bui National Park

Building the Bui Dam affected the national park and its tourism. Maxwell, the Assembly man, and also tour guide, trapper, and supporter of the Bui National Park was willing to answer questions concerning the Bui National Park and the former tourism.

According to Maxwell Gbadego, the tourism on the river started in 1990. Two or three tourists could sit in a canoe, which was moved by a guide who had a long paddle to push the canoe upstream the Black Volta in shallow water. The tour guides knew where the hippos were hiding and brought the tourists near the pools, where the hippos were bathing. Up to 50 hippos could have been observed in only 1 pool, of which there were 9 available for tourists. The indigenous guides were sitting on sandy islands, climbing on mountains and slept in treehouses with the tourists to educate and inform them about their culture, environment, and of course the protected hippos. To the question if the BPA was thinking about the biodiversity and the unique environment of the national park when planning the dam, Maxwell denied. There are things like history and culture, which cannot get restored and which will forever be a loss for the park. One example are the hippos. To resettle the hippos downstream of the dam would have been the easiest way and could have saved their lives and also the tourist attraction. He proposed to the BPA to resettle the hippos himself and to bring them to places where they could find their original habitat as well as being secure. The BPA, on the other hand, reported bringing people from South Africa with special equipment to resettle the hippos. Maxwell considered that carrying these huge animals with special boats or cranes would be against the nature of the animals. Moreover, hippos need shallow water and sandy islands, which is not found upstream the dam in the reservoir. Another risk to resettle them upstream is that the hippos might leave the country and will be killed by people living upstream in Burkina Faso or Ivory Coast, where hippos are not protected animals.

4. DISCUSSION

In order to improve the livelihood of the affected people, it is necessary to listen to their needs. It seems like that one big issue represents the insufficient compensation of farmland, which was given by the BPA. Especially women suffer a lot.

What people need is sustainable help from the BPA on various matters to restore and improve their livelihood themselves. They do not want money from the BPA as a means of compensation for the harm suffered. What they need are fertile farmland with sustainable irrigation systems, good fishing grounds, and new job opportunities for women and the youth to achieve their aims. Having knowledge in farming and fishing is what they have; what is missing, however, are the chances and resources to implement their skills. A sustainable and longer-term assistance of the BPA or other NGOs is required and unavoidable.

If the BPA could help the affected farmers with an improved and sustainable irrigation system, farmers could successfully farm and boost the local markets again. Fishers also suggest a modern way of fishing with ecological aqua cultures downstream the dam to achieve a higher amount of fishes. As a result of these changes, the high unemployment rate of women could be reduced as well, as women are the ones trading goods at the market. The new roads and transport facilities could help to implement the creation of new markets to move their goods outside the villages to have access to new business opportunities. That could be a great chance for women to not depend on the income of their husbands any longer. With the profits, parents could ensure a proper education for their children and could achieve a better life for them. A good education could also help to prevent young girls from getting married and giving birth too early. Upgrading the area with more schools, nurseries, and health centers will further help to improve the living conditions as well as creating new job opportunities and training centers, especially for the women in the villages.

Moreover, it turned out that there is a high improved sanitation backlog demand in all the villages. On the one hand, sanitation facilities are simply missing or not sufficient and on the other hand, the few KVIPs they have are not well placed. Moreover, the toilets in the houses, which the BPA provided in the new resettlements, are not constructed well and smell. The reason for the smell could be that there is a problem with the ventilation and air supply or the shallow depth of the septic tank dump of the toilet. This leads to the fact that many people are still going to the bush, without knowing the serious effects this could bring. Pathogenic bacteria in human excreta can leach into the groundwater, which is pumped from the boreholes in the communities. Drinking contaminated water from the boreholes can lead to a spread of diseases in the communities, which shows how important improved sanitation systems are.

However, it can be said that the water situation in the new resettlements and also in Bongase and Jama Host Community has improved compared to the situation before the dam was constructed. The only problem is the small number of boreholes, which are too few with regard to the number of people in the villages. Part of the compensation agreements was an increased number of boreholes for each community, which has not been completely fulfilled until now.

It is anticipated that in the near future the BPA will restore the tourism. Ecotourism, water activities on the lake, and adventure trips through the park could be really lucrative in the future. Learning about hippos, watching them, and appreciating the unique environment they are living in will help to protect the animals and tourism could help to reach this aim. Maxwell Gbadego also sees the financial benefits of the tourism. The Bui National Park and the hippos could attract many people who will come and pay just to see the environment and habitats of the hippos and other animals. This will also lead to the creation of new job opportunities for the local people. "Looking at the lake behind the dam and seeing the beauty of nature inspires me of new tourist attractions."

5. CONCLUSION

The results of the questionnaires indicate that the impacts of the Bui Dam affected the livelihood of most of the people in a negative way. Resettled people who lost everything as well as those people who lost their source of income, such as farmlands or fishing grounds, have not yet recovered from the loss of their livelihood. Fishing, farming, and trading are the major activities to ensure the livelihood for these people, which have been mainly destroyed. Compensation of the BPA should be used to restore and improve the lives of the affected people. However, most of the compensation has not been sufficient and people still await some of the promised compensation from the BPA. The main issue is the new farmland given to some of the resettlers. The soil is not as fertile as it used to be before, the farms are far away, and the harvest is not enough for trading at the market. Also, the fishers complain about the new fishing grounds, which are not abundant anymore with fishes, since the dam was constructed. Especially women and children are suffering as a result of the loss of livelihood and the unemployment. Women's activities like trading and farming are no more of any need, since there is less to trade at the market. Missing income has led to the fact that women depend on their husbands, which is a regression of the emancipation of rural women there. Missing job opportunities and a bleak future outlook, especially for the youth, is expected to cause an extensive emigration from the area.

Furthermore, the number of KVIPs, health care facilities, as well as schools is no longer sufficient to reach an improved livelihood. The affected people are also unsatisfied with the missing participation of the planning process of the dam, whereby the importance of community participation was clearly stated by the RPF. However, none of the people I spoke to had any influence on the construction and the compensation agreements. The BPA neither knew which kind of compensation they want, nor did they listen to their needs and their concerns. Building the dam also affected the Bui National Park and flooded about 21% of the park's area. Habitats of hippos and other exotic animals have been destroyed and these animals have been displaced. Also, the tourism of the national park is paralyzed due to the fact that the hippos are gone. It remains to be hoped that the BPA will meet the announcements written in the ESMP. "Further consideration will be given to demarcating and gazetting an "offset area" of land to be incorporated into the Bui National Park, that encompasses riverine gallery forest" (ERM, 2007, p. 27).

Even though the construction of the Bui Dam achieved some benefits in the study area, such as better access to drinking water, improved roads, and better accommodations, the negative impacts outweigh its benefits for the local communities. Moreover, the possibilities women hoped for, such as selling food at the construction site for the workers, did not bring the expected benefits.

It needs a better understanding of the socioeconomic impacts of dam constructions to gain more knowledge of large dams and their benefits, especially for vulnerable groups like women and children.

REFERENCES

Barry, B., Obuobie, E., Andreini, M., Andah, W., Pluquet, M., 2005. The Volta River Basin: Comprehensive Assessment of Water Management in Agriculture. Comparative Study of River Basin Development Management.

BPA, 2015. www.buipower.com. http://buipower.com/. (Retrieved from 20 August 2016).

ERM, 2007. Environmental and Social Management Plan (ESMP) for the Bui Hydropower Project.

Hensengerth, O., 2011. Interaction of Chinese Institutions with Host Governments in Dam Construction—The Bui dam in Ghana (D. I. Entwicklungspolitik, Ed.) Bonn.

Hildyard, N., 2008. Dams on the Rocks: The Flawed Economics of Large Hydroelectric Dams. Corner House, Dorset.

Miine, L.K., 2014. Sustainability of Bui Resettlement Scheme in Ghana. (Kumasi Ghana).

Modern Ghana, 2015. www.modernghana.com: http://www.modernghana.com/news/599416/1/bpa-curtails-black-flymenace-at-bui.html (Retrieved from 27 August 2016).

Raschid, L.-S., Koranteng, R., Kyei Akoto-Danso, E., 2008. Research, Development and Capacity Building for the Sustainability of Dam Development with Special Reference to the Bui Dam Project. (Ghana).

Tornyie, F., 2015. Bui hydroelectric power dam project in Ghana. EJOLT Factsheet No. 25.

UCS, 2013. Union of Concerned Scientists. www.ucsusa.org. http://www.ucsusa.org/clean_energy/our-energy-choices/renewable-energy/how-hydroelectric-energy.html#.V6mTfNSLSt9. (Retrieved from 29 July 2016).

UNEP, 2010. Africa Water Atlas. Nairobi, Kenya.

WCD, 2000. Dams and Development—A New Framework for Decision-Making. Earthscan Publications Ltd., London.

WHO, 2015. Dracunculiasis. www.who.int. http://www.who.int/mediacentre/factsheets/fs359/en/ (Retrieved from 26 August 2016).

Wyse, S., 2011. Snap Surveys. http://www.snapsurveys.com/blog/what-is-the-difference-between-qualitative-research-and-quantitative-research/. (Retrieved from 3 September 2016)

Zoomlion Ghana Limited, 2016. www.zoomlionghana.com. http://www.zoomlionghana.com (Retrieved from 26 August 2016).

Chapter 10

Peri-urban Households' Constraints to Water Security and Changing Economic Needs Under Climate Variability in Ghana

Divine O. Appiah, Patrick Benebere, Felix Asante
Kwame Nkrumah University of Science and Technology, Kumasi, Ghana

Chapter Outline

1. **Introduction**   137
2. **Methodology**   139
   2.1 Profile of the Study Area   139
   2.2 Research Design   140
3. **Results and Discussion**   143
   3.1 Changing Economic Needs   143
   3.2 Local People's Perception
       of Climate Variability   146

3.3 Adaptation to Climate
    Variability   151
3.4 Constraints for Adaptation
    to Water Insecurity Under
    Climate Variability   154
4. **Conclusion and Recommendation**   155
Acknowledgments   156
References   156

1. INTRODUCTION

Adaptation to the climate variability and climate change in respect of water insecurity have emerged as a solution to addressing its impacts that are already evident in some regions. This is particularly considered relevant for the Wa municipality, where households are already struggling to meet the challenges posed by existing climate variability (Yamin et al., 2005), and are therefore expected to be the most adversely affected by climate variability and change (McCarthy et al., 2001).

Despite the significance attached to adaptation, there remains a lack of understanding of the key constraints that impede the effective implementation of adaptation strategies by households across Sub-Saharan Africa (SSA) (Antwi-Agyei

Sustainable Hydropower in West Africa. https://doi.org/10.1016/B978-0-12-813016-2.00010-1

137

et al., 2013). Constraints are defined as factors, conditions, or obstacles that reduce the effectiveness of adaptation strategies (Huang et al., 2011). Smit and Pilifosova (2001) identified a list of determinants of adaptive capacity and when not sufficiently available act as constraints. They grouped them into six classes: (i) economic resources, (ii) technology, (iii) information and skills, (iv) infrastructure, (v) institutions, and (vi) equity. The IPCC (2007) remained rather general about constraints to adaptation, defining them as obstacles to reaching a potential that can be overcome by a policy, program, or measure. Moser and Ekstrom (2010) showed that a broad range of constraints can come about in different phases of the adaptation process, from understanding the adaptation problem to planning appropriate adaptation measures until finally managing the planned measures and monitoring the outcomes. They arrived at a conclusion that policies, programs, and measures designed to overcome constraints to adaptation need to be highly context specific, i.e., to consider the respective system of interest, the spatial context, and the actors involved for each phase of the adaptation process.

Constraints and barriers have been distinguished from limits to climate adaptations. In the fourth assessment report of the Intergovernmental Panel on Climate Change (IPCC), limits are defined as conditions or factors that render adaptation ineffective as a response to climate change and are largely insurmountable (Adger et al., 2007). Limits to climate adaptation are endogenous and absolute and therefore unsurpassable (Dow et al., 2013). Many frameworks and approaches have been developed to understand the limits and constraints to climate adaptations (Lehmann et al., 2013). Adaptation to climate change have been constrained by many factors including economic resources (poverty and economic status), technology development and dissemination of information and skills, governance structure (legal, policy, and institutional), sociocultural perspectives, environmental and health issues, and conflicts among different interest groups (Burton et al., 2001).

The IPCC (2001 cited by Ngigi, 2003) stated that lack of technology has the potential to seriously impede a community's ability to implement adaptation options by limiting the range of possible responses and interventions. Adaptive strategy and capacity is likely to vary, depending on availability and access to technology in all sectors (Adger and Brooks, 2003). Many of the adaptive strategies for managing climate change directly or indirectly involve technology. Hence, a community's level of technology and its ability to adapt and modify technologies are important determinants of adaptive capacity. Awareness of and sensitization to the development and utilization of new technologies are also key to strengthening adaptive capacity (Chapman et al., 2004). In many cases, technology choices are limited by inadequate financial resources and knowledge. Technology is costly, so water users must either make money to use it or receive subsidies or incentives to adopt it. Technology development and dissemination are other concerns associated with low adaption. Slow adaptation in Africa can be attributed to low-technology adoption; thus, enhanced water users' education would hasten technology dissemination and climate change adaptation (Dinar et al., 2008). Inadequate funds, technical skills, and capacities affect promotion and adoption of appropriate technologies. Capacity building through demonstration, exchange visits, and incorporation of sociocultural aspects are paramount to any technology transfer package. Technology dissemination or project implementation should embrace participatory and multisectoral approaches to ensure effective stakeholder involvement and sustainability (Ngigi, 2009).

Poverty is directly related to vulnerability, and is therefore a rough indicator of the ability to cope and adapt (IPCC, 2001 cited by Ngigi, 2003). Adaptation of new technology costs money, and because poor communities have less diverse and more restricted entitlements, they lack the empowerment to adapt, locking them into a vulnerable situation. In many parts of Africa, according to Lawrence et al. (2002), the poverty index is directly proportional to water availability, especially for farming communities. Even exploitation of water resources for economic gains is hindered by poverty. In effect, poverty is a major hindrance to adaptation to climate change. According to the Wa municipal profile, poverty is endemic especially among the rural and peri-urban people who are mostly peasant farmers (UK Essays, 2013). This situation makes it difficult for the people to meet the cost of most utilities including water. The situation also has the potential of disenabling them from acquiring or making use of available technologies that can serve as adaptation to climate variability and change.

Extensive research on climate adaptation has been conducted (Ford et al., 2011, Gifford, 2011), yet the majority of studies do not report on constraints to adaption action. Thus, the study of constraints to adaptation and of ways to overcome them is gaining critical interest in adaptation research. The main focus of the chapter is to outline peri-urban households' constraints for adaptation to water insecurity caused largely by climate variability and change in the Wa municipality.

2. METHODOLOGY

2.1 Profile of the Study Area

The Wa Municipal Assembly is the only Municipality in the Upper West Region. There are however 10 additional District Assemblies in the region. Wa Municipal Assembly lies within latitudes 1°40′ N to 2°45′ N and longitudes 9°32′ W to 10°20′ W. It shares administrative boundaries with the Nadowli District to the North, Wa East District to the east, Wa West District to the West, and by the Tuna Kalba District in the Northern Region to the South. The Municipal Assembly has its capital as Wa, which also serves as the regional capital. It has a landmass area of approximately 234.74 km^2.

The climate of the municipality is typical of tropical continental. The area experiences two seasons: the dry and the wet seasons. The wet season lasts between June and September. The dry season starts in late October when the weather is cold and temperatures could be as low as 15°C at night. This extends into the month of March when the weather is extremely hot with dry hazy winds and a maximum temperature of 45°C in the day. The annual rainfall ranges between 840 mm and 1400 mm, and this has serious implications for food crop production and the availability of both surface and underground water sources. The rainfall pattern in this part of the country is as erratic as in most part of the north (Logah et al., 2013). This rainfall scarcity renders the area relatively dry, leading to low access to water in any form. Much has been done in providing water for the Upper West Region for various applications, but the problem still persists. Currently, there is a project under construction to pipe water from the Black Volta to supplement the water needs of the Wa municipality. However, significant climate variability within the region can affect the sustainability of water in the Black Volta. Fig. 1 is a map illustrating areas selected for the study within the Wa municipality.

According to the 2010 Population and Housing Census (PHC), the Wa municipality had a total population of 107,214 with Wa town alone constituting about 65% of the total population (Ghana Statistical Service, 2010). By implication, there is a high density of population in Wa and consequently pressure on water resources, land, and socioeconomic infrastructure. This has extended to the peri-urban areas and raised issues of population pressure specifically on housing, water management, conflict management, land-use planning as developmental issues to be grappled with.

2.2 Research Design

The cross-sectional survey was used in collecting both quantitative and qualitative data through the use of questionnaire and interviews. This was relatively quick and easy to conduct and also allows data to be collected on all variables once, bringing out multiple outcomes and exposures in the study. This design was chosen because it involved a systematic approach to data collection and presentation to reflect a given situation within the period of the study. Thus, variables relevant to the study were gathered from a cross section of targeted communities to achieve the objectives of the study.

2.2.1 Sampling and Sampling Procedure

The study population included some 20 communities and a staff from the CWSA in the Wa municipality. According to the database of the Wa Municipal Assembly, these 20 communities have homogeneous characteristics of water-related challenges within the municipality. The communities are *Kperisi, Boli, Sing, Konta, Kunfabiala, Zingu, Nakori, Dobile, Kpongu, Siru, Sombo, Chakor, Dukpong, Charia, Bamaho, Gurumuni, Mangu, Kumpuhu, Dunku, and Kpongu.* The staff of the CWSA was purposively selected. A total of 345 household respondents were interviewed from 7 communities (representing one-third of the 20 communities) randomly selected from these 20 communities. According

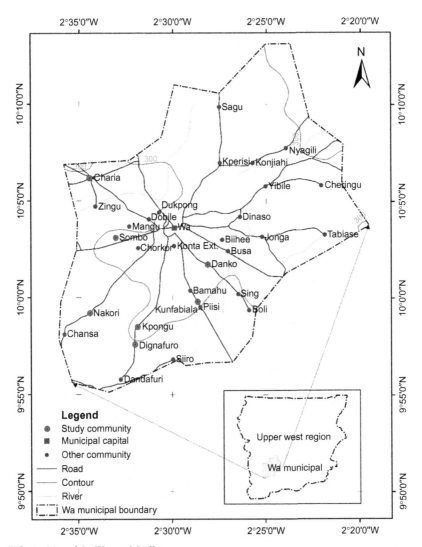

FIG. 1   Map of the Wa municipality.

to Sarantakos (1998), a population is well represented by one-third of the entire population. The names of the communities were written and folded into a container, mixed thoroughly, and seven were randomly picked from among the rest. The rationale for using a simple random sampling technique was to ensure that all communities had a fair chance of being selected. The sample sizes for the various communities were arrived at by the use of proportionate sampling. Table 1 illustrates the sample size of the various communities that were selected.

TABLE 1 Name of Community, Number of Households, and Questionnaire Allocated

Name of Community	No. of Households	Sample Size Per Community
Sombo	973	134
Charia	522	72
Kpongu	408	56
Danko	266	37
Nakori	201	28
Kunfabiela	69	10
Dignafuro	61	8
Total	2500	345

Within each community, 60% of the questionnaires were allotted to women. Women compared to men suffer most in times of water crises, so it was appropriate that more of their views on hindrances to adaptation were taken. The systematic sampling technique was applied in selecting the houses where household respondents were to be drawn from. The value of the total number of houses was used in dividing by the total sample size of each community to arrive at the interval (i.e., the Kth number) between the houses from which households were to be drawn. This was repeated for the rest of the communities. The households in each community that were readily available and willing to respond to the questionnaires were the ones the questionnaires were administered to while taking into consideration the allocations of questionnaires given to the various sex groups.

2.2.2 Data Collection Tools

This chapter made use of both primary and secondary data sources. The primary data were collected from the various households. The secondary data were obtained from institutions such as the Community Water and Sanitation Agency (CWSA) and the Wa Municipal Assembly (WMA). Specific data-gathering tools included an interview guide and a questionnaire.

The household questionnaires were designed to contain both closed and open-ended questions. Items in this research instrument included the local people's perception of climate variability, mode of adaptation to climate variability, and constraints for adaptation to water insecurity under climate variability. The rationale for the use of the questionnaires was to cover many respondents within a short time period and to provide broad information about the issues being studied. Two research assistants were used to administer the questionnaire to the respondents in their local language.

The interview guide was used in interviewing a staff of the Municipal Community Water and Sanitation Agency. Items that were included in the guide were based on the current water situation within the municipality, the types of water facilities that are provided, the reliability of these facilities in the face of climate variability, and measures to improve adequate water supply. The questions that were asked during the interview were open ended. This enabled the interviewee express himself to the fullest to help the study obtain accurate and necessary information.

2.2.3 Data Analysis

Using the Statistical Product for Service Solution (SPSS) version 20, the quantitative data were analyzed and the results presented employing descriptive statistics. The qualitative data were analyzed based on content analysis. Pictorial presentations were also used to support both the quantitative and the qualitative data findings.

3. RESULTS AND DISCUSSION

3.1 Changing Economic Needs

From the data gathered, only 42% out of the 345 respondents were engaged in economic activities. According to Grey and Sadoff (2007), water is an input to almost all production processes and has no substitute. It was the pivot to various economic activities embarked on by the respondents. The other 52% respondents who were not engaged in any economic activity was as a result of their inability to access water. The economic activities included *pitoo* brewing (local beer), agroprocessing, rice milling, irrigation farming, and food vending (see Fig. 2).

The demand for water for the various activities varies from activity to activity. But the quanta of the activities are not static and often not measured. This made it difficult quantifying the amount of water needed for the various activities. During the lean season when water scarcity becomes severe, most of the people are often put out of business. The poorest of the poor are usually the most affected by lack of access to water for productive purposes, resulting in a vicious cycle of malnutrition, poverty, and ill health (Rijsberman, 2006). Pitoo brewing was practiced in Sombo, Charia, Kpongu, Kunfabiela, and Dignafuro. It was not practiced in Danko and Nakori because these two communities were Muslim-dominated communities where alcohol is mostly not consumed. Sixty-seven percent of the respondents who were into pitoo brewing said water was insufficient for the activity. Shea butter processing was also common in all the communities studied except in Charia and Danko. Water was insufficient for 54% of the respondents who were into shea butter processing.

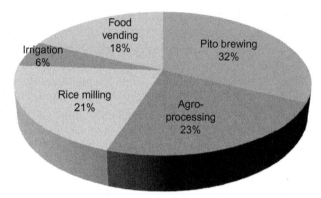

FIG. 2 Economic activities. *(Data from Field Survey, 2015.)*

Water insufficiency for shea butter processing was particularly much an issue for respondents in Kpongu and Nakori as compared to the other communities. Rice milling was common in Kpongu, Danko, Nakori, and Dignafuro. Water was insufficient for 60% of the respondents engaged in the activity. Irrigation farming was practiced in Sombo and Charia and all the respondents who were engaged in activity indicated that water was insufficient. Sullivan et al. (2003) indicated that inadequate water availability is a cause-and-effect of poverty, and these effects exacerbate the poverty trap. In the Charia community, the availability of a well-constructed dam and the many boreholes including other water sources made it possible for them to engage in irrigation activities. In the Sombo community, those who were able to afford water from the small water system to supplement their insufficient boreholes enabled them to engage in irrigation activities. For food vending, it was common in Sombo, Charia, Kpongu, Danko, and Kunfabiela. Out of the 30 respondents, 93% indicated that water was quite sufficient for their activities (Table 2).

From the municipal profile, agricultural activities dominate any other economic activity. Majority of the people (about 70%) were engaged in crop farming activities with the rearing of livestock on a small scale. The sale of the farm produce and animals helps them to pay their water levies and also meet other household expenses. Notwithstanding, the municipality has not been able to diversify its agricultural activities due to the unimodal and erratic rainfall pattern that does not favor farming activities all year round.

Agricultural production is the primary source of direct and indirect employment and income for most rural populations in many developing countries (Ringler, 2007). It is the major sustainer of most African economies and it contributes substantially to the gross domestic product (GDP) and remains the largest source of foreign exchange. It is also the main generator of savings and tax revenues (Ludi, 2009). Dry season gardening through irrigation becomes an important supplement to the normal farming season. However, high temperatures resulting in intensive evaporation often cause the drying up of water bodies

TABLE 2 Water Sufficiency for Economic Activities

WS	Pitoo Brewing		Shea Butter Processing		Rice Milling		Irrigation		Food Vending		Total	
	Count	%	Count	%	Count	%	Count	%	Count	%	Count	%
VS	0	0	0	0	5	14	0	0	0	0	5	3
QS	17	33	17	46	9	26	0	0	28	93	71	43
NS	35	67	20	54	21	60	10	100	2	7	88	54
Total	52	100	37	100	35	100	10	100	30	100	164	100

NS, not sufficient; QS, quite sufficient; VS, very sufficient; WS, water sufficiency.
Data from Field Survey, 2015.

with onset of the dry season. This situation does not favor dry season farming. Improvements in the agricultural sector by way of adapting to climate variability and change can maximize rural incomes and increase the purchasing power for many people and assist bridging the poverty gap (Wiggins, 2006).

The data gathered revealed that all the respondents in Kpongu, Danko, Kunfabiela, and Dignafuro were not engaged in dry season farming because of insufficient water availability. The other respondents who engaged in dry season farming in the Sombo, Chaire, and Nakori communities cultivated common vegetables such as pepper, tomatoes, okra, pumpkin, beans leaves, garden eggs, and bitter leaves on a small scale for household consumption. Dry season farming is seriously hampered by inadequate water availability following inadequate rainfall. Farmers often obtained water from boreholes and dugouts for their crops. These dugouts often easily dry up and also the increasing population mounts intense pressure on the few boreholes, making dry season farming almost impossible.

These climatic threats that decrease agricultural productivity are changing the demographic characteristics of the municipality. Many of the rural folks are migrating to other parts of the country especially the southern parts, ostensibly in search of employment opportunities. According to Ray (2011), lack of opportunities for gainful employment in villages is leading to an ever-increasing movement of poor families to the cities for better economic opportunities. The constant migration of the youth normally results in the neglect of families and also lack of labor for farm work and other activities. This according to the respondents affect their farming activities. Consequently, poverty levels in the region are well above the national figure of 24.2% (GSS, 2012).

As a measure to contain the situation and ensure availability of water all year round, wells have been dug on river beds in some communities but some of these wells dry up during the dry season, indicating a possible reduction in ground waters. As others are trying to contain the situation, some respondents have already thrown their arms up in despair and claim they are waiting on their God while others suggested government should provide more irrigation dams to enable them conserve water for dry season farming.

3.2 Local People's Perception of Climate Variability

About 98% of the respondents could provide evidence of climate variability and change within their environment. These evidence included loss of the forest vegetation comparing the current vegetation with the vegetation 20 years ago, low and patchy rainfall, delayed commencement of the farming period, low crop yield (A respondent lamented that, "*if it were those years, by this time, there would have been new yam in the market, but this is August and there is no new yam, so there isn't much to talk about changes in our time*".), excessive sunshine and heat intensity, increase in numerous diseases including the recent outbreak of Ebola. They however could not associate these changes to any concept. This was is consistent with a conclusion by Gyampoh et al. (2009) that the indigenous people may not understand the concept of global warming or climate change but they rightly observe and feel its effects: decreasing rainfall, increasing air temperature, increasing sunshine intensity, seasonal changes in rainfall patterns, and massive crop failure.

The respondents attributed crop failure in the municipality to low rainfall, prolonged drought, and changes in rainfall pattern. Agriculture in the municipality is rain-fed and farmers have over the years perfected the art of predicting the onset of the rainy season. A respondent in a focus group discussion said, "*...the flying past of certain black birds towards the north normally marked the beginning of rainy season and when these birds fly back towards the south it indicated the start of the dry season. These birds no longer fly past as they used to*". Predicting the onset of the rains has however become difficult in recent years due to significant changes in the rainfall pattern. This was found to be consistent with a claim by Logah et al. (2013) that recent climate trends show highly variable interannual and interdecadal rainfall amounts in Ghana making it very difficult to predict with certainty the long-term trends of rainfall in the country. The beginning of the rainy season is no longer predictable and sometimes during the rainy season, there may be an unexpected long break before the rains resume. This situation alongside increase in sunshine and heat intensity causes the wilting of crops.

Plate 1 depicts the nature of some farmlands at the onset of drought during the rainy season. This situation makes it difficult for farmers to plan their cropping seasons to coincide with the rains to ensure maximum crop yield. Also prolonged droughts cause a reduction in the water available in the soil for crop growth. The result is crop failure and low crop yield. When this occurs, money spent on land preparation and planting, as well as expected income from the sale of farm produce is lost and household savings is normally spent to replant farms. Respondents' perceptions about the nature of climatic elements within the municipality are presented in Fig. 3.

PLATE 1   A farmland in Charia during an onset of drought.

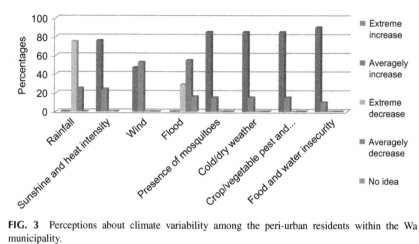

FIG. 3 Perceptions about climate variability among the peri-urban residents within the Wa municipality.

The respondents attributed these changes to several factors. About 43% claimed the changes are as a result of their bad farming practices such as bush burning and indiscriminate cutting down of trees. Also about 32% of the respondents believed that the changes regarding the various climatic elements are a command from God to punish them for their evil deeds especially on low rainfall. As stated in the words of one of the discussants in a focus group discussion, *"We have been entrusted with the power to call rain in periods of drought but Christians and Muslims preach against us and describe us as people who like food. As a result we have decided to stop so that all of us will suffer. After all we will all face the difficulties"*. However about 25% of them could not associate the changes noticed to anything as they had no knowledge about the factors causing the changes.

The study also observed that most of the people in many of the communities harvest firewood and engage in commercial charcoal production. Others also fell trees to harvest their leaves to sell to herdsmen as fodder for their herds during the dry season. These practices account for excessive deforestation that can drive the climate to become variable (Plate 2).

Despite the enormous indicators of climate variability and change being noticed, about 81% of the respondents indicated their ignorance about climate variability or change terminology. This is possibly due to their low level of education. A crosstabulation between their educational status and knowledge of climate variability or change was carried out and is displayed in Table 3.

From Table 3, people with higher educational status had high knowledge on climate variability than those with low educational status. However, the data revealed that there is a high rate of illiteracy among the peri-urban people within the Wa municipality, therefore, majority of them representing about 81% had no knowledge of climate variability. A chi-square test for significance between respondents' educational background and their awareness of climate variability was statistically significant at $P < .000$. This is illustrated in Table 4.

PLATE 2 Abandoned charcoal production mound at Dignafuro.

TABLE 3 A crosstabulation of the Educational Status of Respondents and Their Knowledge of Climate Variability or Change

Educational Status of Respondent	Knowledge of Climate Variability or Change		Total (%)
	Yes (%)	No (%)	
No education	14.2	85.8	100.0
Primary school	5.9	94.1	100.0
Junior high school	30.4	69.6	100.0
Senior high school and above	52.9	47.1	100.0
Total	19.4	80.6	100.0

The educational background and the awareness of climate variability are directly related. The fact that majority of the respondents do not have formal education has the tendency of limiting their ability to choose from the list of appropriate technological options that are available and can be used to minimize the impact of climate variability on their livelihoods. As suggested by Lee (2007), successful adaptation requires knowledge about available options, the capacity to access the options, and the ability to choose and implement the most suitable ones. This requires capacity building in the form of providing formal education. People with low educational status are more likely to lack knowledge on climate change and may not be able to adapt; therefore, if the educational levels of the people are low, then it is expected that their ability to adapt to climate variability and change will also be low.

The language spoken by the respondents did not have any direct translation of climate terminology and this added to their lack of knowledge on climate variability and change terminology. This supports the claim by Ungar (2000)

TABLE 4 Chi-Square Tests on Educational Background of Respondents and Their Awareness of Climate Variability

| | Value | df | Asymp. Sig. (2-Sided) | Monte Carlo Sig. (2-Sided) | | | Monte Carlo Sig. (1-Sided) | | |
| | | | | | 95% Confidence Interval | | | 95% Confidence Interval | |
				Sig.	Lower Bound	Upper Bound	Sig.	Lower Bound	Upper Bound
Pearson chi-square	38.198[a]	3	.000	.000[b]	.000	.009			
Likelihood ratio	34.257	3	.000	.000[b]	.000	.009			
Fisher's exact test	33.819			.000[b]	.000	.009			
Linear-by-linear association	26.387[c]	1	.000	.000[b]	.000	.009	.000[b]	.000	.009
No. of valid cases	345								

[a] 0 cells (0.0%) have expected count less than 5. The minimum expected count is 6.60.
[b] Explains the significance of the association at 0.001.
[c] Explains the value of the association for which all tests were significant.

that climate terminology does not have standard translation in local languages in Ghana and therefore make Ghanaians more aware that their climate has been changing than their understanding of climate terminology. In a focus group discussion, the respondents were asked to describe some characteristics of climate variability in using at least a word or a phrase.

These characteristics included an increase in the occurrence of drought, rainstorm, erosion, flood, and crop failures. These occurrences were together described as a change of the weather by some of the respondents, while others also described the happenings as the occurrence of a disaster. This is coherent with the conclusion by Neville and Anu (2010) that many people are unfamiliar with climate variability, climate change, global warming, and related terms. It also affirms the assertion by Weber (2010) who suggested that average individuals seem to be typically more concerned about the weather rather than the climate in their respective regions.

3.3   Adaptation to Climate Variability

The data gathered revealed that none of the respondents had ever attended any sensitization workshop on climate variability or change. It was therefore not surprising when about 84% of the respondents claimed they were vulnerable to climate variability as a result of their ignorance about the terminology of climate variability and change. Chapman et al. (2004) argued that awareness of and sensitization to the development and utilization of new technologies are also key to strengthening adaptive capacity. This is not achievable without knowledge of the issue that needs adaptation. People first need to recognize an issue that requires them to adapt to it. As claimed by Doss and Morris (2001), people first need to recognize that their climate has changed before suggesting an adaptation strategy. This will tune their minds to thinking about a possible adaptation mechanism. All these are achievable through capacity building in the form of formal education, workshops, and sensitization programs. A crosstabulation between their educational background and their mode of adaptation to climate variability was carried out and is displayed in Table 5.

It is observed from Table 4 that 63% of the respondents are those with no idea of adapting to climate variability. This is to be expected because people without formal education are more likely to have no knowledge about climate variability and hence might not also have knowledge on possible adaptation measures. As suggested by Nelson et al. (2007), building adaptive capacity increases the ability of individuals, groups, or organizations to transform the adaptive capacity into action. Also Doss and Morris (2001) assert that if indigenous people are to adopt water conservation techniques, they must first be aware that the technology exists and perceive that it is profitable. This is achievable through building adaptive capacity. A chi-square test for significance between the respondents' educational background and their mode of adaptation was statistically significant at $P^< .000$. This is illustrated in Table 6.

TABLE 5 A Crosstabulation of the Educational Status of Respondents Versus Their Mode of Adaptation to Climate Variability

Educational Status	Engaging in Other Economic Activities (%)	Planting Trees (%)	No Idea (%)	Relying on God (%)	Educating People (%)	Total (%)
No education	52.3	62.3	63.9	57.1	0.0	59.1
Primary school	16.3	14.8	13.3	20.0	0.0	14.8
Junior high school	20.9	8.2	15.2	22.9	20.0	16.2
Senior high school and above	10.5	14.8	7.6	0.0	80.0	9.9
Total	100.0	100.0	100.0	100.0	100.0	100.0

TABLE 6 Chi-Square Tests on Education and Mode of Adaptation to Climate Variability

	Value	df	Asymp. Sig. (2-Sided)	Monte Carlo Sig. (2-Sided)			Monte Carlo Sig. (1-Sided)		
				Sig.	95% Confidence Interval		Sig.	95% Confidence Interval	
					Lower Bound	Upper Bound		Lower Bound	Upper Bound
Pearson chi-square	41.564[a]	12	.000	.000[b]	.000	.009			
Likelihood ratio	33.641	12	.001	.000[b]	.000	.009			
Fisher's exact test	29.074			.000[b]	.000	.009			
Linear-by-linear association	.054[c]	1	.817	.829[b]	.789	.869	.435[b]	.382	.487
No. of valid cases	345								

[a]5 cells (25.0%) have expected count less than 5. The minimum expected count is .49.
[b]Explains the significance of the association at 0.001.
[c]Explains the value of the association for which all tests were significant at 0.05.

The educational background of the respondents and their mode of adaptation to climate variability are directly related. People with high educational status are more likely to be able to adapt to climate variability than people with low educational status or without formal education. Capacity building is therefore an important component of any climate change adaptation strategy.

Regardless of the respondents' vulnerability to climate variability, they still had ways of adapting to it. About 28% of them indicated they try to engage in more than one economic activity so that if one fails due to climate variability, the other might succeed. Respondents from *Sombo*, *Charia*, *Kunfabiela*, and *Dignafuro* communities were those basically into such alternative practices. Some added food vending, including the preparation of local beans cake called *koose* to the brewing of *pitoo*, a local beer; but their challenge was how to raise capital for these economic ventures. Nevertheless, how could this mechanism minimize climate variability or water insecurity? This kind of adaptive mechanism to climate variability and water insecurity is rather not tenable because *pitoo* brewing and frying *koose* all require the use of water, which might not be readily available at all times due to climate variability.

However, these adaptive mechanisms were suggested and practiced because they were ill-informed about climate variability. One would have expected to hear about adaptation practices geared toward protecting the various water bodies such that the water in them could last for a long time to support the various economic activities embarked upon. This was not to be because they indicated their vulnerability. This chapter will want to agree with a suggestion by Schipper (2007) that work on adaptation should not only be focused on addressing the impacts of climate change, but rather be extended to addressing the underlying factors that cause vulnerability to it.

Only 19% of the respondents from *Sombo* and *Charia* communities practiced tree planting either around their houses or on their farms. This adaptive mechanism was good but also needed to have been practiced around their water bodies to reduce the rate of evaporation. Majority of the respondents representing about 54% were those who did not have any way of adapting to climate variability. For them, climate variability is a punishment from God for their wrongful doings.

3.4 Constraints for Adaptation to Water Insecurity Under Climate Variability

The respondents had different adaptation strategies to water insecurity at the household level. One such strategy was rainwater harvesting. This supported the other insufficient water sources (Liuzzo et al., 2016). This was very necessary following periods of prolonged drought leading to the drying up of rivers and streams that were expected to have provided water all year round. However in all the communities covered by the study, not enough rainfall was harvested. This was because of the inadequate rainfall in terms of quantity and distribution, as well as poor harvesting technology (Morey et al., 2016).

Most of the respondents depended on the roofs of their buildings to harvest rainwater. However, the roofing system, which is predominantly thatch, could not support efficient rainwater harvesting, since provisions were not often made to channel rainwater into various reservoirs as per the roofing system. Only a small percentage of the rainwater could be harvested with majority going untapped. This was the case because they lack the financial strength to improve upon their housing system and hence the rainwater harvesting technology. This is consistent with Ekstrom and Moser (2014), who identified inadequate financial and political resources as the common constraints encountered by local people in trying to adapt to climate variability. This situation constrained the respondents' ability to adapt to water insecurity.

As part of the adaptation processes to ensure the availability of water all year round, wells have been dug on river beds in some communities, even so, some of these wells dry up during the dry season, indicating a possible reduction in ground waters that probably occur as a result of excessive rise in temperature and evaporation coupled with low rainfall in the area.

The provision of mechanized boreholes by the municipal assembly and some private individuals was meant to also add to the adaptation process to water insecurity at the household levels among the peri-urban residents within the Wa municipality. This, however, came with its own challenges as the respondents are expected to pay for electricity consumed before obtaining water. According to the municipal profile, poverty is prevalent especially among the rural and peri-urban people who are mostly peasant farmers. This situation makes it difficult for the people to meet the cost of most utilities including water. According to the respondents, they wish to have their normal boreholes from which they used to draw water for free. This situation occurred because the municipal assembly failed to consult broadly with the communities before providing them the water facilities.

Adaptation strategies are expected to reflect the needs and aspirations of the society or community it is meant to benefit (Ngigi, 2009). This was however not the case with this adaptive strategy. Most of those who could not afford the bills look elsewhere for water but with health implications as these sources are unprotected dug outs and rivers. And normally when these sources dry up, it poses a very serious challenge to them. This has contributed to the water insecurity situation among the peri-urban people within the Wa municipality. These situations have come about because the Municipal Assembly failed to broadly consult with the communities before mechanizing the boreholes.

4. CONCLUSION AND RECOMMENDATION

The chapter espoused that people in the Wa Municipal Assembly, like many people throughout Ghana and beyond, are experiencing the impact of climate variability through water stress; however, the same cannot be said about their level of awareness of the scourge, in order to fashion out appropriate adaptation measures. Climate variability coupled with population growth in the peri-urban areas within the municipality also affected the availability of water resources. Inadequate knowledge on climate variability and overdependence on climate sensitive resources made the local people vulnerable to climate variability.

The people have adapted to climate variability in various ways. However, the adaptation strategies contribute very little to the minimization of their woes. The chapter also observed that low existence of technology to harvest rainfall, climate variability, and poverty were the constraints for adapting to water insecurity and climate variability.

The chapter recommends that public education and awareness campaigns should be undertaken by the Municipality and other relevant stakeholders to improve knowledge of the issues as well and to deepen understanding on the relationship between water resources and climate variability and change in the Municipality. This will ameliorate the excessive anthropogenic activities that contribute to the occurrence of climate variability and climate change. The Wa Municipal Assembly in collaboration with Nongovernmental organizations (NGOs) should provide sustainable alternative sources of water supply other than rainwater to the residents in the peri-urban areas. More so knowledge on the requisite technology needed for rainwater harvesting should be made available to the residents of the peri-urban communities by various stakeholders. This will enable them store enough rainwater for use in the lean season. Again, climate change terminologies should be translated into local languages by experts, for easy understanding by the local people.

ACKNOWLEDGMENTS

We acknowledge the CWSA within the municipality and the Wa Municipal Assembly (WMA) for making available information to our disposal that supported the success of this chapter. We thank greatly the household respondents for their time in availing information during the fielding of the household questionnaires. Finally, we acknowledge the secondary sources cited as well as thank the anonymous reviewers of this manuscript for their constructive critiques on the chapter.

REFERENCES

Adger, W.N., Agrawala, S., Mirza, M.M.Q., Conde, C., O'brien, K., Pulhin, J., Pulwarty, R., Smit, B., Takahashi, K., 2007. Assessment of Adaptation Practices, Options, Constraints and Capacity. Cambridge University Press, Cambridge.

Adger, W.N., Brooks, N., 2003. Does Global Environmental Change Cause Vulnerability to Disaster? Routledge, London.

Antwi-Agyei, P., Dougill, A.J., Stringer, L.C., 2013. Barriers to Climate Change Adaptation in Sub Saharan Africa: Evidence from Northeast Ghana and Systematic Literature Review. Sustainability Research Institute (SRI) University of Leeds, Leeds, United Kingdom.

Burton, I., Challenger, B., Huq, S., Klein, R.J.T., Yohe, G., 2001. Adaptation to Climate Change in the Context of Sustainable Development and Equity. Cambridge University Press, UK.

Chapman, R., Slaymaker, T., Young, J., 2004. Livelihoods Approaches to Information and Communication in Support of Rural Poverty Elimination and Food Security: The Literature Update. Overseas Development Institute (ODI). Available from www.fao.org/rdd/doc/SPISSL-LiteratureUpdate.pdf (Accessed 3 November 2013).

Dinar, A., Hassan, R., Mendelsohn, R., Benhin, J., 2008. Climate Change and Agriculture in Africa: Impact Assessment and Adaptation Strategies. Earthscan, London.

Doss, C., Morris, M., 2001. How does gender affect the adoption of agricultural innovations? The case of improved maize technology in Ghana. Agric. Econ. 25 (4), 27–39.

Dow, K., Berkhout, F., Preston, B.L., Klein, R.J., Midgley, G., Shaw, M.R., 2013. Limits to adaptation. Nat. Clim. Chang. 3 (4), 305–307.

Ekstrom, J.A., Moser, S.C., 2014. Identifying and overcoming barriers in urban climate adaptation: case study findings from the San Francisco Bay Area, California, USA. Urban Clim. 9, 54–74.

Ford, J.D., Berrang-Ford, L., Paterson, J., 2011. A systematic review of observed climate change adaptation in developed nations. Climate Change 10 (6), 327–336.

Ghana Statistical Service, 2010. 2010 population and housing census. . Summary of final report.

Gifford, R., 2011. The dragons of inaction: Psychological barriers that limit climate change mitigation and adaptation. Am. Psychol. 66 (4), 290–302.

Grey, D., Sadoff, C.W., 2007. Sink or swim? Water security for growth and development. Water Policy 9 (2), 545–571.

GSS, 2012. The 2010 population & housing census summary report of final results. A publication of the Ghana Statistical Services, May, 2012, 117 p.

Gyampoh, B.A., Amisah, S., Idinoba, M., 2009. Using Traditional Knowledge to Cope with Climate Change in Rural Ghana. Faculty of Renewable Natural Resources, Kwame Nkrumah University of Science and Technology (KNUST), Kumasi.

Huang, C., Vaneckova, P., Wang, X., Fitzgerald, G., Guo, Y., Tong, S., 2011. Constraints and barriers to public health adaptation to climate change: a review of the literature. Am. J. Prev. Med. 40 (2), 183–190.

Intergovernmental Panel on Climate Change, 2001. Climate Change 2001: Impacts, Adaptation and Vulnerability. Cambridge University Press, Cambridge.

Intergovernmental Panel on Climate Change, 2007. IPCC 3rd assessment report—climate change 2001: working group II: impacts, adaptation and vulnerability. . Available from http://www.ipcc.ch/ipccreports/index.htm (Accessed 19 August 2013).

Lawrence, P., Meigh, J., Sullivan, C., 2002. The Water Poverty Index: An International Comparison. Keele Economics Research Papers, KERP, No. 19, Keele Economics Department, Keele University, Keele, Staffordshire. October.

Lee, B.L., 2007. Information technology and decision support system for on-farm applications to cope effectively with agrometeorological risks and uncertainties. Available from http://www.springerlink.com/content/n07834253472x7q4/ (Accessed 19 August 2013).

Lehmann, P., Brenck, M., Gebhardt, O., Schaller, S., Subbauer, E., 2013. Barriers and opportunities for urban adaptation planning: Analytical framework and evidence from cities in Latin America and Germany. Mitig. Adapt. Strateg. Glob. Chang. 8, 1–23.

Liuzzo, L., Notaro, V., Freni, G., 2016. A reliability analysis of a rainfall harvesting system in southern Italy. Water 8 (1). 18 pp.

Logah, F.Y., Obuobie, E., Ofori, D., Kankam-Yeboah, K., 2013. Analysis of rainfall variability in Ghana. Latest Research in Engineering and Computing 1 (1), 1–8.

Ludi, E., 2009. Climate Change, Water and Food Security. Overseas Development Institute, London.

McCarthy, J.J., Canziani, O.F., Leary, N.A., Dokken, D.J., White, K.S., 2001. Impacts, Adaptation, and Vulnerability to Climate Change. Cambridge University Press, Cambridge.

Morey, A., Dhurve, B., Haste, V., Wasnik, B., 2016. Rain water harvesting system. Int. Res. J. Eng. Technol. 3 (4), 2158–2162.

Moser, S.C., Ekstrom, J.A., 2010. A framework to diagnose barriers to climate change adaptation. Proc. Natl. Acad. Sci. U. S. A. 107 (51), 22026–22031.

Nelson, D.R., Adger, W.N., Brown, K., 2007. Adaptation to Environmental Change: Contributions of a Resilience Framework. Available from http://environ.annualreviews.org (Accessed 12 September 2013).

Neville, L., Anu, M., 2010. Ghana talks climate; the public understanding of climate change. Available from www.africatalksclimate.com (Accessed 27 November 2013).

Ngigi, S.N., 2003. Rainwater Harvesting for Improved Food Security: Promising Technologies in the Greater Horn of Africa. Greater Horn of Africa Rainwater Partnership (GHARP), Kenya Rainwater Association (KRA), Nairobi.

Ngigi, S.N., 2009. Climate Change Adaptation Strategies: Water Resources Management Options for Smallholder Farming Systems in sub-Saharan Africa. The MDG Centre for East and Southern Africa at the Earth Institute at Columbia University, New York.

Ray, S., 2011. An analysis of the trend in economic capacity utilization and productivity growth of some energy intensive industries in India: 1979–80 to 2003–04. Econ. Bull. 31 (4), 55.

Rijsberman, F.R., 2006. Water scarcity: fact or fiction? Agric. Water Manag. 80 (1–3), 5–22.

Ringler, C., 2007. The Impact of Climate Variability and Climate Change on Water and Food Outcomes: A Framework for Analysis. International Food Policy Research Institute. Global Assessments: Bridging Scales and Linking to Policy, GWSP Issues in Global Water System Research No. 2.

Sarantakos, S., 1998. Social Research, second ed. Macmillan, Basingstoke.

Schipper, E.L.F., 2007. Climate Change Adaptation and Development: Exploring the Linkages. Tyndall Centre Working Paper No. 107 20.

Smit, B., Pilifosova, O., 2001. Adaptation to climate change in the context of sustainable development and equity. In: McCarthy, J.J., Canziani, O., Leary, N.A., Dokken, D.J., White, K.S. (Eds.), Climate Change 2001: Impacts, Adaptation and Vulnerability. IPCC Working Group II. Cambridge University Press, Cambridge, pp. 877–891.

Sullivan, C.A., Meigh, J.R., Giacomello, A.M., Fediw, T., Lawrence, P., Samad, M., Mlote, S., Hutton, C., Allan, J.A., Schulze, R.E., Dlamini, D.J.M., Cosgrove, W.J., Delli Priscoli, J., Gleick, P., Smout, I., Cobbing, J., Calow, R., Hunt, C., Hussain, A., Acreman, M.C., King, J., Malomo, S., Tate, E.L., O'Regan, D., Milner, S., Steyl, I., 2003. The water poverty index: development and application at the community scale. Nat. Resour. Forum 27 (4), 189–199.

UK Essays, 2013. November Community Profile of the Wa Municipality . Retrieved from: https://www.ukessays.com/essays/tourism/community-profile-of-the-wa-municipality.php?cref=1.

Ungar, S., 2000. Knowledge, ignorance and the popular culture: climate change versus the ozone hole. Climate Change 9 (3), 297–312.

Weber, E.U., 2010. What shapes perceptions of climate change? Climate Change 1 (3), 332–342.

Wiggins, S., 2006. Agricultural Growth and Poverty Reduction: A Scoping Study. Working paper no. 2 on globalization, growth and poverty, IDRC, Ottawa.

Yamin, F., Rahman, A., Huq, S., 2005. Vulnerability, adaptation and climate disasters: a conceptual overview. IDS Bull. 36 (4), 1–14.

Chapter 11

Legislation on Hydropower Use and Development

Lilian Idiaghe
University of Ibadan, Ibadan, Nigeria

Chapter Outline

1. Introduction	159	4.3 Sierra Leone	175
2. The International Landscape	162	4.4 Liberia	176
3. Hydropower in the ECOWAS Region	165	4.5 Gambia	178
		4.6 Cape Verde	180
4. National Frameworks for Hydropower Use and Development	168	4.7 Mali	181
		4.8 Senegal	183
4.1 Ghana	168	5. Conclusion and Recommendations	184
4.2 Nigeria	171	References	186

1. INTRODUCTION

Climate change is presently a global challenge that impacts countries irrespective of natural boundaries. It has been defined as a change of climate, which is attributed directly or indirectly to human activity that alters the composition of the global atmosphere and is in addition to natural climate variability observed over comparable time periods.[1] The journey to the realization of climate change as a complex challenge and its inclusion in the international agenda began at the United Nations Conference on the Human Environment in Stockholm in 1972. The Conference was organized to address the environmental harm occasioned by transboundary pollution and hazardous waste. Within its narrow concept, it produced a Declaration of 21 principles—"The Stockholm Declaration" (United Nations, 1972). The Declaration is considered as an international marker—as the first international Treaty to recognize the interrelatedness of the human environment and development and to set the pace for what is today referred to as international environmental law. Although not a legally binding Treaty, the Declaration significantly set the structure of global interaction about

---

1. Article 1, United Nations Framework Convention on Climate Change.

Sustainable Hydropower in West Africa. https://doi.org/10.1016/B978-0-12-813016-2.00011-3

development and the environment. Specifically, Principle 21 of the Declaration captured the need for actions on environmental protection to thrive beyond national boundaries. The Principle stated that:

> *States have, in accordance with the Charter of the United Nations and the principles of international law, the sovereign right to exploit their own resources pursuant to their own environmental policies, and the responsibility to ensure that activities within their jurisdiction or control do not cause damage to the environment of other States or of areas beyond the limits of national jurisdiction.*

The Declaration called for multilateral measures[2] to address the effect of human activities on the environment and was reiterated by the Report of the World Commission on Environment and Development 15 years later. Recent global concern about the environment, sustainable development, and climate change can be traced back to the work of the Commission.[3] The Commission's Report (World Commission on Environment and Development, 1987) defined sustainable development as "development that meets the needs of the present without compromising the ability of future generations to meet their needs." It articulated that to achieve such development, the global energy mix needed to undergo substantial changes. Within this mix, it mapped out a clear role for up-to-date policies, laws, and institutional cooperation required to achieve the changes and as well as the need to close the huge gap between international law and national laws. It further identified renewable energy including hydropower as the untapped potential and a key element of the solutions toward sustainable energy. Notwithstanding, the relevance of the Commission's recommendations for a sustainable future, climate change did not become prominent in the field of energy until the United Nations Framework Convention on Climate Change (UNFCCC)[4] was adopted in 1992 as one of the outcomes of the United Nations Conference on Environment and Development (UNCED). The Conference was, in part, a result of the Brundtland Report because the debates and Conference Documents were framed from the Report. As a framework Convention, UNFCCC allows protocol to be adopted from it and the Kyoto Protocol[5] was adopted pursuant to the Convention in 1997.

In 1988, the World Meteorological Organization (WMO) and the United Nations Environment Program (UNEP) established an Intergovernmental Panel on Climate Change (IPCC) to provide the world with a clear scientific view on the current state of knowledge in climate change and its potential environmental and socioeconomic impacts.[6] The panel also reviews and assesses the

---

2. Principle 24.
3. Also commonly referred to as the Brundtland Commission after its Chairman Gro Harlem Brundtland, a former Norwegian Prime Minister.
4. http://unfccc.int/resource/docs/convkp/conveng.pdf (Retrieved 30 June 2016).
5. http://unfccc.int/resource/docs/convkp/kpeng.pdf (Retrieved 30 June 2016).
6. https://www.ipcc.ch/organization/organization.shtml (Retrieved 3 July 2016).

most recent scientific, technical, and socioeconomic information produced worldwide relevant to the understanding of climate change. The IPCC relies on peer-reviewed and published scientific/technical literature to review technical and scientific data relevant to understanding climate change risks, and measures for mitigation and adaptation. The Panel's first Report was published in 1990 and was the basis for climate change concerns to form part of the agenda of the Conference and eventual adoption of the UNFCCC.

The IPCC works through three Working Groups (WGs), a Task Force, and a Task Group. WG I assesses the physical scientific aspects of the climate system and climate change. WG II assesses the vulnerability of socioeconomic and natural systems to climate change, consequences of and options for adapting to climate change. WG III assesses options for mitigating climate change through limiting or preventing greenhouse gas emissions and enhancing activities that remove them from the atmosphere.[6]

Most recently, the United Nations Sustainable Energy for All (UNSE4ALL) Initiative was launched in 2011 to ensure universal access to modern energy services, double the global rate of improvement in energy efficiency, and double the share of renewable energy in the global energy mix by the year 2030. It requires each country to prepare and implement an Action Plan that strategically maps out the policy, institutional, financial infrastructural, and human capital structure to enable each country achieve these goals. It is important to understand the basic underpinnings of international and regional efforts toward the development of hydropower in order to understand how countries can replicate or domesticate internationally recognized rules and principles.

Against the background that countries must make decisions based on their own circumstances, the chapter also examines the regional policy and regulatory efforts to promote the use and development of hydropower in West Africa. Accordingly, at regional level, the Economic Community of West African States (ECOWAS) Treaty, Action Plans and Policies and the other regional instruments emanating from the Community including the West African Power Pool (WAPP) will be examined in assessing the efforts made by West African countries toward the development and use of hydropower in their respective states. This is in an attempt to highlight the initiatives of the regional body to garner partnerships and understanding among member States toward the development of the region's energy sector as it relates to hydropower.

The commitment and efforts of the Economic Community of West African States (ECOWAS) that portend regional and multilateral cooperation is essential to drive the region's goal of universal access by 2030, but national commitments vary. Hence, the national measures are also assessed alongside the commitment to implementation of regional and international agreements that have the capacity to accelerate hydropower use and development.

The persuasive nature of international agreements and principles, and the practice of dualism by most countries in Africa in that internationally agreed principles do not have automotive application to the national home front indicate that hydropower legislation mostly falls within the residual powers of national governments even though these will be influenced by regional and international trends to various extents. It is apt to underscore that international and regional efforts have rapidly increased in the last decade and continue to expand rapidly, indicating that they may very well be the real cutting edge of hydropower legislation in the future. The Brundtland Commission also indicated that interstate cooperation in hydropower development could transform supply potential in neighboring developing countries in Africa.[7]

This chapter assesses the legislative landscape for hydropower use and development in West Africa. The reason that legislative measures are so significant is that the future progress in hydropower deployment is significant to the development of a sustainable energy framework for the region's future. The development of energy resources and advancing their significant inclusion in the energy mix largely depends on the availability and strength of government policies, and laws as well as the institutional feasibility to implement such policies and laws. This portends a tripartite structure comprising policies, laws, and institutions, where the policies create implementing legal frameworks that in turn establish strong institutions. These are further examined at the international, regional, and national levels.

2. THE INTERNATIONAL LANDSCAPE

The Stockholm Declaration, without express mention of climate change, called for cooperative spirit by all countries, big and small, on an equal footing to handle international matters concerning the protection and improvement of the environment.[8] It emphasized that cooperation through multilateral or bilateral arrangements or other appropriate means is essential to effectively control, prevent, reduce, and eliminate adverse environmental effects resulting from activities conducted in all spheres. The Declaration stated that such cooperation should be in such a way that account is taken of the sovereignty and interests of all States. Given the different ways through which sovereignty of States has been eroded, for example through multilateral and bilateral treaties and globalization, the "equal footing" for all States is an unrealistic assumption. The demarcation between countries under the Kyoto protocol, for example, underscores the disparity among the capacity or ability of countries to cooperate as equals.

However, at regional level, the cooperation on equal footing is more amenable to adoption. For example, all ECOWAS countries are member of the United Nations, signatories to both the UNFCCC and the Kyoto Protocol. They are also on the list of Least Developed Countries (LDCs) except Nigeria, Ghana,

7. Our Common Future, http://www.un-documents.net/our-common-future.pdf (Retrieved 3 July 2016).
8. Principle 24.

and Ivory Coast. Furthermore, countries can take a regional position on international platforms on climate change or hydropower development. Within the context of hydropower, the regional landscape of the Economic Community of West African States (ECOWAS) was adopted to understand how legislation affects hydropower development and climate.

The Rio Declaration on Environment and Development emerged as one of the outcomes of the United Nations Conference on Environment and Development (UNCED) held in Rio de Janeiro, Brazil, in 1992. Signatory countries made a critical statement to achieve a balance between environmental protection and development. In other words, countries committed not to undertake development activities at the expense of the environment. Albeit a nonbinding Treaty, it provides principles that countries can draw on to approach development without impeding environmental protection and their natural resource conservation.

Hydropower is a low-carbon renewable energy resource and has been proven as a better choice for providing clean, low-cost electricity, but not without likely negative environmental impacts such as destruction of ecosystems and loss of biological diversity. Hydropower in this way poses as a solution and a challenge at the same time. By seeking a balance between development and environment, the Declaration placed hydropower use and development in critical perspective. The Plan of Implementation of the World Summit on Sustainable Development 2002 outlined that the global drive to increase the contribution of renewable energy throughout the world must include hydropower development. In recent times, the sustainability of hydropower projects necessarily requires inclusion of environmental and social impact assessment, indicating that sustainable development is a core consideration for hydropower use and development. All West African countries were represented at the Conference, but the Declaration is not binding and commitment to implementing its principles at national level is not water tight due to its persuasive nature. Generally, while governments may need to commit to sustainability principles in hydropower development, Principle 8 specifically states that unsustainable patterns of production and consumption can secure a higher quality of life for all people; this is particularly applicable in the context of the energy sector.

The UNFCCC requires all State Parties to formulate, implement, publish, and regularly update national, and where appropriate, regional programmes containing measures to facilitate adequate adaptation to climate change.[9] A unique feature of the Kyoto Protocol is that the countries classified under Annex B agreed to reduce greenhouse gas emissions 1990 levels. The Protocol employed the principle of "'common but differentiated responsibilities" to place a heavier burden on developed nations or Annex I countries. The 43 countries in Annex I,[10] which are mainly industrialized countries and economies in transition, agreed to individually or jointly reduce their overall emissions of such gases by at least

---

9. Article 4 (1) (b).
10. Also referred to as Annex I countries.

5% below 1990 levels in the commitment period 2008–12. Demonstrable progress toward their achievement of commitments under Protocol was to be seen by 2005.[11] Non-Annex I countries are mostly developing countries and were not obliged to undertake any immediate restrictions, but to rely on funding and technology transfer from Annex I countries.

The UNFCCC created an international reporting structure that consists of communication of information related to implementation,[12] settlement of disputes,[13] and annual meetings of the Conference of the Parties.[14] The options for addressing climate change have been divided into four categories: emissions' control, energy efficiency and conservation, long-term carbon storage or carbon sequestration, and adaptation (Dernbach and Kakade, 2008). Hydropower is the largest renewable source of electricity generation and as such hydropower schemes raise climate change concerns. As a result, hydropower use and development intersects electricity generation legislation and climate change policy. This follows the realization that the development of hydropower has implications for climate change and as such climate change policy and hydropower policy must also intersect.

The Sustainable Energy for All initiative (SE4ALL) was launched in September 2011 by UN Secretary-General Ban Ki-Moon to make sustainable energy for all a reality by the year 2030. As a follow up, the United Nations General Assembly declared 2012 as the International Year of Sustainable Energy for All and the year 2014–24 as the International Decade of Sustainable Energy for All. These will contribute to enabling Member States and the UN system to increase the awareness of the importance of addressing energy issues and to promote action at the local, national, regional, and international levels. The Initiative has three core objectives: (i) ensuring universal access to modern energy services; (ii) doubling the rate of improvements in energy efficiency; and (iii) doubling the share of renewable energy in the global energy mix.[15] The Initiative is hinged on the idea that Access to modern energy services is fundamental to human existence.

The adoption of the Sustainable Development Goals (SDGs) in September 2015 to succeed the Millennium Development Goals (MDGs) as the global sustainable development agenda till 2030 and the adoption of the Paris Agreement by consensus by all of the 196 UNFCCC Parties (195 States and 1 regional economic integration organization) have renewed the need for West African countries to critically consider and take actions toward climate change in keeping with an international development agenda. The challenge for many of these countries lies in domesticating these international obligations to ensure that

---

11. Article 3.
12. Article 12.
13. Article 14.
14. Article 7.
15. http://www.se4all.org (Retrieved 20 June 2016).

domestic efforts are in sync with global requirements and vice versa. Of unique importance are the Intended Nationally Determined Contributions (INDCs) of each country. The persuasive nature of international law notwithstanding, it has significantly shaped the debates, policy options, and commitment to climate change mitigation and adaptation. Implementing the INDCs at the national level will be a good first step in the preferred direction.

Notwithstanding, the requirement of domestication or specific adoption by national legislature will still need to be satisfied for dualist States. This is so because even though the INDCs emanated from national government, they are incorporated into the Paris Agreement and as such would need domestication of the Agreement to enable their implementation. As of October 2016, Ghana has taken the lead in this direction. In keeping with the provisions of Article 75 of the Constitution that requires any treaty, agreement, or convention executed by or under the Authority of the President in the name of Ghana is made subject to ratification either by an Act of Parliament or by a resolution of Parliament supported by the votes of more than one-half of all the Members of Parliament, the Parliament of Ghana ratified the Paris Agreement on Climate Change in August 2016.

3.   HYDROPOWER IN THE ECOWAS REGION

In 1999, the West African Power Pool (WAPP) was created by Decision A/DEC.5/12/99[16] of the 22nd Summit of the Authority of ECOWAS Heads of State and Government to integrate the operations of national power systems into a unified regional electricity market, which will, over the medium to long term, assure the citizens of ECOWAS Member States of a stable and reliable electricity supply at competitive cost. The Pool is projected to develop through three consecutive phases. Phase 1 commenced in 2012 with the commissioning of the regional transmission infrastructure, trading arrangements, and establishment of the regulator. Phase 2 proceeds from the previous phase and the third phase reflects a long-term vision to optimize the operations of the pool. As of 2014, hydropower made up 31% of current electricity generation in the region (Adeyemo, 2014). The Pool will provide the essential building block for a sustainable energy infrastructure network in the region. The mandate of WAPP includes the development of large hydropower projects of 100MW and above. This leaves less priority for small-scale hydropower projects. Considering the goal of WAPP to facilitate crossborder energy trade between utilities in the member countries, small-scale projects could be directed to local consumers and increasing access to electricity in rural areas. All ECOWAS States are members of WAPP.

An Environmental Policy for ECOWAS was adopted in 2008 at the 35th ordinary session of the Authority of Heads of State and Government Abuja, Nigeria.

---

16. http://www.ecowapp.org/?page_id=10 (Retrieved 24 June 2016).

The Policy covers all aspects of natural resources management, environmental protection, and human settlements, particularly of the urban habitat. It was aimed at reversal of the state of degradation of natural resources and to improve the quality of their living conditions and environment, to conserve biological diversity, so as to secure a healthy and productive environment by improving the ecosystem balance and the well being of the populations.[17] As part of its strategic lines of action, the policy called for the set up and support for a functional regional technical consultation to monitor and boost the implementation of the international Conventions with special attention to the UNFCCC and the Kyoto protocol. Notwithstanding, ECOWAS does not have a climate change policy and no mention of climate change or hydropower development is made in the Policy. It remains unclear how the policy can implicitly impact hydropower development as a strategy to address the challenge of climate change occasions by high carbon emissions.

Most of the regional efforts toward the development of hydropower in the region have been initiated under the auspices of the Economic Commission of West African States (ECOWAS). These efforts are consistent with the ECOWAS Treaty (Economic Community of West African States, 1975), which required the development of renewable energy in the region. The various ECOWAS energy policies and initiatives are traceable to Article 28 of the Treaty, which states that:

1. Member States shall coordinate and harmonize their policies and programmes in the field of energy.
2. To this end, they shall:
    a. ensure the effective development of the energy resources of the region;
    b. establish appropriate cooperation mechanisms with a view to ensuring a regular supply of hydrocarbons;
    c. promote the development of new and renewable energy particularly solar;
    d. include energy in the framework of the policy of diversification of sources of energy;
    e. harmonize their national energy development plans by ensuring particularly the interconnection of electricity distribution networks.[18]

ECOWAS Renewable Energy Policy[19] was adopted in 2012 in keeping with the Treaty as noted here. The purpose of the Policy is to enable increased use of renewable energy resources for grid electricity supply and for the provision of access to energy services in rural areas. It was designed to complement

---

17. Article 5.
18. http://www.ecowas.int/wp-content/uploads/2015/01/Revised-treaty.pdf (Retrieved 2 July 2015).
19. Economic Community of West African States (ECOWAS) Renewable Energy Policy 2012. http://www.ecreee.org/sites/default/files/documents/basic_page/151012_ecowas_renewable_energy_policy_final.pdf (Retrieved 13 July 2015).

conventional sources for power production like large hydro and natural gas. The policy primarily targets the electricity sector, but also considers some additional issues regarding the domestic energy sector. The renewable energy Directive of the European Union provides a strong model for the policy as the former indicated that regional cooperation is essential to garner national commitment toward a common objective. The commitment of ECOWAS Member States to mainstream renewable energy into their national policies requires cooperation in order to surmount the multifaceted barriers they currently encounter from national initiatives.

The policy marks the initial steps of aligning the national renewable energy policies with the ECOWAS renewable energy (EREP). The plan of the policy was to translate same into national targets and activities by developing National Renewable Energy Policies (NREPs) and corresponding Action Plans by the end of 2014. The Policy aims to enable universal access to electricity in the region by 2030. The NREAPs are 5-year bound and their implementation of NREAPs is coordinated and monitored by ECREEE on behalf of the ECOWAS Commission.

The ECOWAS White paper of 2006 was adopted as a regional policy to support access to energy services for rural areas known by ECOWAS Heads of States. The White Paper aims to contribute to the Millennium Development Goals (MDG) and to reduce poverty. This effort to improve energy access for populations in rural and periurban areas in the region forecasts that at least 20% of new investment in electricity generation should originate from locally available renewable resources. These would in turn ensure that member States are able to achieve self-sufficiency and sustainable environmental development. ECOWAS Centre for Renewable Energy and Energy Efficiency (ECREEE) in Cape Verde in 2010 are key steps to establish the institutional feasibility necessary to support small hydropower development in the region. ECREE drives renewable energy in the region through technological capacity and facilitating exchanges between Member States.

The ECOWAS Small-Scale Hydropower Program (First Phase: 2013–18) jointly developed and executed by ECREEE and United Nations Industrial Development Organization was adopted by the ECOWAS Ministers of Energy in October 2012 to be implemented between 2013 and 2018. The program was aimed at establishing an enabling environment for small-scale hydropower investments and markets in the ECOWAS region. As a priority action under the regional SE4ALL framework for West Africa, the Small-Scale Hydropower Program (SSHP) contributes to the objective of the ECOWAS Renewable Energy Policy to increase the share of renewable energy (excluding large hydro) in the region's electricity mix to an average 10% in 2020 and 19% in 2030. It will boost the ECOWAS White Paper on Energy Access in Periurban and Rural Areas, which targets 25% renewable energy-based service to rural population. The SSHP complements the West African Power Pool (WAPP) Master Plan, which is mainly focused on the expansion of transmission lines and generation from large hydro and natural gas.[20] The UNIDO-Regional Center for Small Hydropower was established in Nigeria in 2006 to implement the SSHPP.

One of the core objectives of the program is to enable strengthening of policy and regulatory frameworks for small hydropower use and development in Member States. It is projected that by 2018 ECOWAS countries would have improved their legal framework and that SSHP would have become an integral part of ECOWAS/WAPP planning documents. Assessment of the relevant existing policies on SSHP policies and legal and regulatory frameworks is a core activity of the program. The assessment will enable comparison with successful policy models across countries.

4. NATIONAL FRAMEWORKS FOR HYDROPOWER USE AND DEVELOPMENT

There is essentially a difference between large hydro and small hydropower systems; while large hydropower systems date many decades back in the region, interest and commitment to developing small hydropower systems is rather recent and has evolved with the advancement of sustainable development, renewable energy and rural access to electricity. In the countries where hydropower makes up the larger percentage of the energy mix, the legislation on the use and development of hydropower were enacted to establish implementing institutions alongside. In Ghana, for example, the Volta River Authority was established in 1961 to harness the resources of the Volta River to primarily provide electrical energy for Ghana. Similarly, in Nigeria, the Electricity Supply Company (NESCO) commenced operations as an electric utility company in 1929 with the construction of a hydroelectric power station at Kurra near Jos. The Electricity Corporation of Nigeria (ECN) was established in 1951, while the Niger Dams Authority (NDA) was established in 1962 with a mandate to develop the hydropower potentials of the country. The now-defunct National Electric Power Authority (NEPA) was created by government Decree No. 24 of 1972 from the merger of the Electricity Corporation of Nigeria (ECN) and Niger Dams Authority (NDA).

The following examination of the legislative landscape for hydropower in some ECOWAS countries is done bearing in mind that hydropower intersects the environment, energy sector, and climate change.

4.1 Ghana

Hydropower is the major source of Ghana's electricity generation—67% in 2011 and a projected 41% additional capacity by 2020 (Government of Ghana, 2011).

---

20. ECOWAS Small-Scale Hydropower Program (First Phase: 2013 to 2018) Document, ECOWAS Regional Centre for Renewable Energy and Energy Efficiency, Cap Verde.

This indicates that Ghana's energy supply is vulnerable to climate change disruptions occasioned by drought and as such requires resilience and mitigation measures. The Government of Ghana has a national energy policy objective to provide affordable access to electricity to all communities by 2020. It also hopes to become a net exporter of electricity by 2015.[21] The policy also set a goal of 10% from renewable energy in the electricity mix by 2020, but this goal does not include the share of large hydropower plants, which already represents a share of between 60% and 70% in its power generation. The government's policy on small hydropower was stated to include the creation of appropriate incentives and a regulatory framework for its development. The Policy aimed to promote the development of hydropower by improving the fiscal and regulatory framework and incentivizing development of small hydropower. This policy objective was realized through the Renewable Energy Act.

Prior to the enactment of the Renewable energy Law in 2011, the challenge of developing small hydroelectric projects persisted due to the absence of a legal and regulatory framework for renewable energy. The Renewable Energy Act, 2011 (Act 832) aims to promote, develop, manage, utilize, sustain, and ensure adequate supply of RE resources for power and heat and other related purposes. RE as defined by the Act includes wind, solar, hydro, biomass, bio fuel, landfill gas, sewage gas, geothermal energy, and ocean energy. The Law created a feed-in-tariff scheme to further incentivize private sector investment in renewable energy. The feed-in-tariff component will ensure return on investment for independent power providers, including small hydrogenerators in both urban and rural areas. Sequel to the Act, the Energy Commission published the License Manual in 2013 for Service Providers in the Renewable Energy Industry to complement the scanty provisions of the Act with regard to independent generation from renewable sources including hydro.

As part of efforts to meet her obligations under the UNFCCC and as a signatory to the Kyoto Protocol, the government of Ghana restated its commitment to achieve 15% penetration of rural electrification by decentralized renewable by 2015 and 30% by 2020. In its second communication to the UNFCCC (Government of Ghana, 2011), it also outlined policy options for coordination of climate change and the Convention activities with the creation of an implementation framework consisting of relevant government agencies led by the Ministry of Environment, Science and Technology, which serves as the focal institution for climate change activities and host of a functional National Committee on Climate Change. The Committee consists of experts from government Ministries, Universities, Research Institutions, the Private sector, and Nongovernmental Organizations and has a Ministerial directive to, among other things, formulate a National Climate Change Policy, map mitigation, and adaptation strategies and make recommendations for scaling up climate change

---

21. Government of Ghana Energy Policy 2012.

training awareness and planning. The Report highlighted the result of a technology needs assessment conducted in 2006 to prioritize portfolios of technologies in the energy sector. Small and mini hydropower technologies emerged last on the list in order of priority after eight other kinds of technologies. It also identified a number of government agencies and academic institutions that mandate border on climate change and the need to enhance their capacity to ensure higher delivery on their mandates. However, these agencies are not strictly energy sector institutions and raise the possibility of lopsided efforts in the absence of strong coordination with energy sector institutions.

Ghana's Environmental Protecting Act was enacted in 1994 to establish an Environmental Protection Agency. The main functions of the Agency include policy planning, coordination of relevant ministries, and collaboration with relevant stakeholders. It is understandable that the Act was made before the Kyoto Protocol was adopted, but it precedes the UNFCCC. However, the provision of Section 3 of the Act comes in useful when construed constructively. The Section enacts that the Minister may give to the Agency such directives of a general nature as to the policy to be followed by the Agency in the performance of its functions as appears to the minister to be necessary in the public interest. The challenge of climate change sufficiently fits this obligation.

In 2015, Ghana became the first African country to use the Hydropower Sustainability Assessment Protocol. The Protocol is a voluntary scorecard by which the hydropower industry can assess the social and environmental performance of its projects. The Protocol was prepared by the Hydropower Sustainability Assessment Forum (HSAF), a self-selected group of industry representatives, government agencies, financiers, and large NGOs. It allows the assessment of projects at four stages (early stage, preparation, implementation, and operation). Projects are assessed across a range of aspects with scores from 1 to 5. The scorecard is not a standard, but could inform policy and design of appropriate frameworks at country levels. The protocol has been criticized for its lack of independence, substantive weaknesses, and the exclusion of dam-affected people from its preparation and use due to the exclusion of civil society groups. Affected people were excluded from the process.

Ghana has a national energy policy and institutional framework on energy and environment, but did not have a climate change law or national policy on climate change that addresses climate change in its complex and multisectoral dimensions until 2013. The country needed a unified strategy that addresses environmental concerns, hydropower development, and climate change challenges in a cohesive manner. Ghana's National Climate Change Policy was approved in 2013 and was closely followed in 2015 by a National Climate Change Master Plan Action Programmes for Implementation: 2015–2020. The Action Plan has an objective—among several others—to secure adequate water for energy production and food security during all seasons. A priority task in this direction is to construct reservoirs to provide hydropower to rural communities.

The Ministry of Energy provides policy direction for the energy sector in Ghana, and is responsible for setting energy policy, including in the power sector. It exercises supervisory functions over the Energy Commission and the Public Utilities' Regulatory Commission in their roles and technical and economic regulators of the energy sector, respectively. The Energy Commission[22] was established in 1997 to regulate the country's energy sector and to coordinate all policies in the sector. The Commission is responsible for issuing licenses to utilities for electricity and natural gas in Ghana. The Public Utilities' Regulatory Commission (PURC)[23] was also established in the same year as an economic regulator while the Energy Commission is the technical regulator. PURC is responsible for setting electricity tariff. Accordingly, existing hydroelectric power utilities, independent small hydropower projects, and potential projects are within the purview of both regulators. However, it is not clear whether there is a clear line of cooperation among the Environmental Protection Agency and both regulators, and the laws are silent on the point.

4.2   Nigeria

Hydropower is presently the second-largest energy resource for electricity generation in Nigeria. Nigeria's Energy Policy[24] was approved by the federal government in 2003 to provide for optimal utilization and management of the country's diverse energy resources. In addition to developing and harnessing hydropower with particular attention to the development of the mini- and micro-hydropower schemes and their integration into the energy mix, the policy also envisaged private sector and indigenous participation in hydropower in the process. Its major objectives were to increase the percentage contribution of hydroelectricity to the total energy mix and employ mini- and micro-hydropower schemes for rural electrification.

Several strategies were outlined to achieve these objectives including the establishment and maintenance of multilateral agreements to monitor and regulate the use of water in international rivers flowing through the country, increased indigenous participation in hydropower stations planning, design and construction, and active private sector participation. It also sought to ensure that rural electricity boards incorporate small-scale hydropower plants in their development plans and active support for research and development activities for the local adaptation of hydropower plant technologies. In order to scale up electricity generation, the policy aimed to develop other potential sites for hydropower, gas, and coal-fired power plants for electricity generation.

The short-term strategy for hydropower implementation was the constant review and improvement of multilateral agreements on the use of water in international rivers flowing through the country, increased local participation, and

22. Energy Commission of Ghana Act of 1997.
23. Public Utilities and Regulatory Commission Act, 1997 (Act 538).
24. Energy Commission of Nigeria, National Energy Policy, 2003.

creation of appropriate fiscal measures as incentives to indigenous and foreign entrepreneurs for the local production of hydropower plants and accessories. Establishment of appropriate institutional arrangements, regulations, and guidelines for the development of small-scale hydropower plants was also outlined. In the medium term, government would introduce alternative technological options to ease the impact of water shortage on hydropower plants and establishment of mini- and micro-hydropower plants. Encouragement of the widespread construction of mini- and micro-hydropower plants was set as a long-term strategy. In keeping with the policy's goal to undergo a review every 10 years, the Energy Commission of Nigeria commenced review of the policy in late 2013.

A Renewable Energy Master Plan was also drafted by the Commission with technical assistance from the United Nations Development Programme (UNDP) in 2005. The Plan identifies considerable potential for generating small and large hydroenergy in the country. The Plan was reviewed in 2014 to make more concise projections based on the country's commitment to universal access to energy. The plan envisions deployment of small and medium hydropower projects for rural areas and as a strategy to increase the share of renewable in the country's energy mix. As a follow up, the Renewable Energy Policy Guidelines were issued by the Federal Ministry of Power in 2006 with the core objective to expand the role of renewable electricity in sustainable development through effective promotional and regulatory instruments. The Guidelines define renewable electricity to include electricity to include small-, mini-, and micro-hydropower. The definition of hydro as stated in the Renewable Energy Master Plan 2005 was adopted:

i. Small hydro is all hydroelectricity schemes below 30 MW
ii. Mini-hydro is all hydroelectricity schemes below 1 MW
iii. Micro-hydro is all hydroelectricity schemes below 100 kW
iv. Pico-hydro is all hydroelectricity schemes below 1 kW

The policy also aimed to develop regulatory procedures that are sensitive to the peculiarities of renewable energy-based power supply. The Guidelines define renewable energy as the use of energy from a source that does not result in the depletion of the earth's resources whether this is from a central or a local source. "Renewable electricity" refers to electric power obtained from energy sources whose utilization does not result in the depletion of the earth's resources. Renewable electricity also includes energy sources and technologies that have minimal environmental impacts, such as less intrusive hydro and certain biomass combustion. These sources of electricity normally will include hydropower. The policy essentially reiterated the institutional and regulatory framework set out by the Constitution and the Power Sector Reform Act of 2005.

The Electric Power Sector reform Act of 2005 is presently the legal framework for the Nigerian electricity industry. It rests on the National Electric Power Policy of 2001 that laid down the road map for reform of the electricity industry.

The Act also established the Nigerian Electricity Regulatory Commission as the sole regulator of the electricity industry in Nigeria. The Commission is empowered to make regulation in furtherance of the provision of the Act.

Environmental protection and enforcement is the responsibility of the National Environmental Standards and Regulations Enforcement Agency (NESREA). It was established by the National Environmental Standards and Regulations Enforcement Agency Act of 2007, which contains scanty provisions for climate change protection. It enacted that the Agency will enforce compliance with the provisions of such treaties to which Nigeria is a party. In accordance with the Constitution[25] that requires every Treaty to be enacted into law by the National Assembly through a process referred to as specific adoption or domestication, such international treaties must have been domesticated before compliance can be enforced by the Agency. The Agency is principally mandated to enforce compliance with the provisions of international agreements, protocols, conventions, and treaties on the environment. The UNFCCC and the Kyoto Protocol, although ratified by Nigeria, have not been domesticated, meaning that the Agency is not at liberty or under strict obligation to enforce its provisions. This essentially leaves Nigeria without specific climate change regime with which to implement her commitments under both Treaties.

A Bill to establish a National Climate Change Commission for Nigeria was presented to Parliament in 2008. The Bill sought to create a Commission with responsibility management and control of climate change and other related environmental matters. The Commission would be charged with responsibility for the strategic planning and coordination of national policies on climate change and energy in all its ramifications. The Bill is yet to be passed into law.

The Bill empowered the Commission to implement both national and international regimes on climate change, act as Designated National Authority (DNA) for the purpose of implementation of the Kyoto Protocol financial mechanisms, and liaise with the United Nations Framework Convention on Climate. The UNFCCC and the Kyoto Protocol have been ratified by Nigeria but are yet to be domesticated. This portends that even when the Bill is enacted into law, the Commission will be unable to, on the strength of the Act, enforce the provision of international treaties. Therefore, domestication of relevant climate change treaties should precede the Act.

Nigeria's First National Communication (Ministry of Environment of the Federal Republic of Nigeria, 2003) to the UNFCCC was made in 2003, and it noted increased use of renewable resources, consisting of the introduction of small-scale hydroplants as a GHG mitigation strategy in the Energy Sector. The Communication precedes the Power Sector Reform Act of 2005 and the NESREA Act of 2007. A second national communication was made in 2014. The actions to combat climate change outlined in the second communication include deployment of more efficient Hydro Electric Power turbines that can

---

25. Section 12 (1) Constitution of the Federal Republic of Nigeria 1999 (as amended).

be installed on small streams for electricity production. This strategy is of high priority and of medium cost that both government and private sector can implement.

The Federal Government, through the Federal Executive Council (FEC), is at the apex of energy policy in Nigeria. FEC provides overall direction for the development of the electricity industry in Nigeria, in alignment with other national policies and also facilitate their alignment with Nigeria's international obligations, especially on climate change. The required legal and regulatory measures required to support renewable electricity policies are also its responsibility.

The Federal Ministry of Power and Steel is saddled with policy making for the electricity industry, including renewable electricity policy. The Ministry also proposes policy options and recommendations to the Federal Government concerning legislation, policy, and investment on renewable electricity; and Monitors and evaluates the implementation and performance of the policy within governmental agencies and in the electricity market. The Ministry has supervisory oversight over the regulatory institutions in the industry.

The Nigerian Electricity Regulatory Commission (NERC) was established by the Electric Power Sector Reform Act of 2005 and is responsible for independent regulation of the electricity industry and the creation, promotion, and preservation of efficient industry and market structures to ensure the optimal utilization of resource for the provision of electricity services. In respect of renewable electricity, the Commission has the duty to develop simplified licensing procedures for renewable energy investments; a framework for power purchase agreement that ensures access to grid-based renewable electricity; preferential prices for renewable electricity to cover additional costs due to size, technology, location, and the intermittent nature of the particular renewable electricity resource base; and lower licensing charges for renewable electricity licensees. It must be noted that even though the Act empowers the Commission to regulate hydroelectric power generation, transmission, and distribution, the Commission does not enforce environmental protection. The latter is within the purview of the National Environmental Standards Regulation and Enforcement Agency. However, at licensing stage, the Commission is bound to ensure that Environmental Impact Assessment (EIA) of each hydropower project has satisfied the EIA standards as regulated by the Agency. NESREA is under the supervision of the federal Ministry of Environment created in 1999 with overall mandate for protecting Nigeria's environment and the conservation of its natural resources. A Climate Change Unit exists in the Department of Environmental Assessment. The Unit exists solely to coordinate the UNFCCC and the Kyoto Protocol and ensure Nigeria's performance of the parties' obligations under both treaties.

A Rural Electrification Agency has the mandate to extend the national grid, develop isolated and mini-grid systems and renewable energy power generation for rural electrification. The Agency was also established to serve as an implementation agency for renewable electricity and provide a coordinating point for renewable electricity activities among state and federal agencies. One of the core implementation strategies of the Agency is the use of renewable energy resource for off-grid projects including small hydro systems. The responsibility of conducting strategic planning and coordination of national policies in the field of energy in all its ramifications lies with the Energy Commission of Nigeria. The Commission was established by Act 62 of 1979 as amended by Acts 32 of 1988 and 19 of 1989.

While the creation of the CCU is a welcome development, there is need for the Unit to maintain coordination and cohesion with the energy sector institutions with regard to combating climate change and creating adaptation strategies through hydropower development. Nigeria could use some more coordination between the regulatory institutions to improve climate change coordination and environmental protection as well as hydropower development.

4.3   Sierra Leone

Sierra Leone emerged from the throes of a 10-year civil war in 2002. As part of efforts toward reconstruction of the country's major sectors, the Energy Policy[26] was designed in 2004 with funding assistance from the Nations Economic Commission for Africa (UNECA). The major goals of the policy are to ensure adequate modern energy supply to meet the development aspirations of the country on a sustainable basis and affordable electricity throughout the country. The policy also sought to develop indigenous fossil and renewable energy sources for the country, especially utilization of such sources to meet the Millennium Development Goals (MDGs) by 2015. The policy also targeted an increase in access to electricity services to 35% of the population of Sierra Leone by 2015. Specifically, the policy emphasized the Government's commitment to promote the development of mini/micro-hydrosites and other renewable energy technologies through different arrangements including public/private partnerships. In April 2010, the Government announced the National Energy Policy Implementation Strategy, launched at the start of Sierra Leone's first Renewable Energy Week, and set out plans for achieving the goals established in the National Energy Policy. According to the country's second-generation Poverty Reduction Strategy Paper for 2009–11, a key government objective is the provision of a reliable power supply in the country, moving toward a low-carbon energy economy through use of the country's significant hydropower potential.

The Ministry of Energy and Power is responsible for the electricity and water sectors. Its functions include policy formulation, planning, and coordination for the electricity sector including hydro and other renewable energy resources' development. The Ministry is also charged with related to electric power supply, including hydroelectric schemes and, nominally, RE matters related to solar and

---

26. Government of Sierra Leone, National Energy Policy, 2002.

wind energy. The National Power Authority (NPA) was established in 1982 to develop the hydroelectric potential of the country. It has sole responsibility for power generation, transmission, distribution, and supply; including hydropower.

Sierra Leone's Environment Protection Agency Act was enacted in 2008 to establish the Environment Protection Agency with responsibility for environmental management and coordination. A National Secretariat for Climate Change (NSCC) was established in May 2012 under the Environment Protection Agency Sierra Leone (EPA-SL), as the Coordinating Body for climate change activities, programmes, and relevant policy development and implementation.[27] This was in response to the omission in the EPA Act to provide for a department for climate change. The NSCC works with other Departments of the EPA-SL and relevant Government Ministries, Departments, and Agencies, involved in climate-related programmes to carry out its daily functions. Although the Act enables the EPA to, among other things, conduct environmental impact assessments, there is no clear indication of how the development of hydropower will be impacted by the Act or the Agency's work.

Sierra Leone's first Communication (Government of Sierra Leone Ministry of Transport and Aviation, 2007) to the UNFCCC in 2007 proposed institutional strengthening in the country's water resources sector given that development of water resources for hydroelectric power production was still in its formative years. The second communication revealed the result of the technology needs assessment that identified a number of mitigation technology options including small hydropower systems. It outlined the reduction of overall GHGs emissions levels through expansion and development of hydropower and solar power as policy objectives.

Sierra Leone has several potential small hydropower sites. Her membership of WAPP has the potential to enable Sierra Leone harness her potential for significant hydroelectric generation and possibly export electricity in the future.

4.4 Liberia

Liberia's Renewable Energy and Energy Efficiency Policy and Action Plan was adopted in 2008 and is presently implemented by the Centre for Sustainable Energy Technology. Both are targeted at facilitating government support through tax subsidies for renewables and increase investment in off-grid rural electrification through the deployment of renewable energy technologies. In 2009, the National Energy Policy was issued to lay down the country's vision for the energy sector. The policy is aimed toward universal access to modern energy services in the country.

The Ministry of Land, Mines, and Energy has the statutory responsibility for the development of mineral, water, and energy resources of the country and the administration of its lands. The Department consists of the Bureau of

---

27. http://epa-sl.org/index.php/departments/climate-change (Retrieved 15 July 2016).

Hydrocarbons and the Bureau of Energy Technology and Policy Development. The Department regulates the electricity sector and is in charge of linkages with energy-oriented organizations, both state controlled and privately owned. In addition, the department monitors and coordinates the energy sector (both conventional and nonconventional). The Rural and Renewable Energy Agency was established after the adoption of the NEP in 2009 to implement rural electrification to rural areas through grid-based and renewable energy technologies, including micro-hydropower systems. The National Energy Policy proposed that the Department shall have at least three divisions managed by Assistant Directors—the Division of Hydrocarbons, with management responsibility over the petroleum sector; the Division of Electricity and Renewable Energy, to look after the electricity sector and promote the development of renewable energy resources; and the Division of Energy Policy and Planning to look after all cross-cutting issues. However, the Division of Electricity and Renewable Energy is responsible for the formulation of electricity sector policy and plans, including the development and review of policies, quality standards, and master plans for grid, off-grid, and renewable energy investments.

The environment protection and management law of the republic of Liberia was enacted in 2002 to create the Environmental Protection Agency as a monitoring, coordinating, and supervisory authority for the sustainable management of the environment in partnership with other government ministries. The EPA is the successor of the National Environmental Commission of Liberia (NECOLIB) that was first established in 1999. The Act does not make any reference to climate change protection or to renewable energy development as a strategy for environmental protection. The EPA serves as the UNFCCC National Focal Point (NFP) and coordinates the preparation of national greenhouse gas (GHG) inventories and compilation of the National Communication (NC) under the UNFCCC.

Liberia's made her first national communication (Environmental Protection Agency of Liberia, 2013) to the UNFCC in 2013. For the energy sector, the country's targets included reducing greenhouse gas emissions by 10% by 2030, improving energy efficiency by 20% by 2030, and raising the share of renewable energy to 30% of electricity production, and 10% of overall energy consumption by 2030. The Mitigation options to achieve these targets include the promotion of hydroelectricity as in the prewar period, promotion of the use of renewable energy technologies and energy-efficient appliances, declaration of emission standards, and reduction of the losses in the electricity supply system. The nonexistence of technology development and diffusion policy has been identified as a challenge to the development of hydropower. The challenge would be effectively addressed by increased awareness among regulatory authorities, investors, and development partners about the potential contribution of hydropower to the country's economy.

Liberia's membership of WAPP could aid development of the country's huge hydropower potential. Part of the development planned for Liberia under the

WAPP project is the cooperative construction of the Mount Coffee and Mano River hydroelectric dams, with the Ivory Coast and Sierra Leone, respectively. These projects would be connected with the partner countries via 220-kV lines, connecting Mount Coffee to the Danane substation, and the Mano River project with the existing substation at the Bumbuna hydroelectric project.[28]

The major adaptation needs and infinitives are summarized relative to the key vulnerable sectors identified during the consultation process and did not include the energy or the environmental sectors (Environmental Protection Agency of Liberia, 2008). Institutional and regulatory structures will be necessary to integrate climate change into national action and development activities, especially in the light of the fact that the country has identified the development of hydropower as a panacea for the challenge of climate change. The civil conflicts that Liberia witnessed in the 1990s and 2000s had a devastating impact on the country's electricity sector. The implication is that the country will need to take steps to counter the significant risk posed by the country's political history for much-needed private investors.

4.5 Gambia

Gambia's National Energy Policy of 2005 has a long-term aim of maximizing efficient development and utilizing scarce energy resources to support economic development in an environmentally friendly way. It also aims to improve and expand energy systems through private sector partnership, and promote renewable energy through private sector participation. The Policy unequivocally stated the government's intention to provide energy security through subregional and international cooperation. Among the strategies identified to achieve this policy objective is the need to build capacity to effectively participate in the development of the Organization for the Development of The Gambia River Basin's (OMVG) hydroelectric power projects and the interconnection of the electric power grids of the subregion. The Electricity Act was enacted in the same year to set out a framework for the licensing of private generation, transmission, and distribution operators, and to promote private sector participation in the electricity sector. It also opened the electricity generation sector to IPPs. As a follow-up, the Programme for Accelerated Growth and Development (PAGE) 2012–2015 outlined strategies for sustainable investment regime and it identified a scale up of the renewable energy resources for electricity generation, and aggressive switch from fossil fuel to renewable energy in the rural areas as a principal strategy to eradicate poverty.

Overall policy direction for the country's energy sector emanates from the office of the President that directly supervises the Ministry of Energy (MOE). The Department of State for Petroleum, Energy and Mineral Resources

---

28. Renewable Energy & Energy Efficiency Partnership, 2012, Liberia. http://www.reeep.org/liberia-2012 (Retrieved 4 July 2016).

(DoSPEMR) within the Ministry of Energy was created in September 2007 to formulate policies across the energy sector. The Public Utilities Regulatory Authority (PURA) was created by the Public Utility Regulatory Act of 2001 and is tasked with responsibility of regulating the electricity industry including the issuance of licenses and safety standards and tariff setting and review. In addition to implementing the regulatory framework, the functions of the PURA entail formulation of the feed-in-tariff price structure and management of the renewable energy fund. The RE fund objective is to provide financial resources for the promotion of RE by providing financial incentives, FITs, capital subsidies, production-based subsidies, and equity participation. These are in the bid to support investment in on-grid and off-grid renewable energy investments to scale up electricity access in remote areas and islands.

In order to enhance capacity to include a larger share of renewable energy in the country's energy mix, a Renewable Energy Bill was drafted by the ministry with the active participation of the PURA and technical assistance from development partners in 2012. This was closely followed by an Electricity Strategy and Action Plan in the same year. While the Plan mainly consisted of propositions for crossborder trade, renewable generation and reliability, and high RE integration as major highlights of the Plan, it also recommended regional interconnection and a joint hydropower project with neighboring countries. This allows for greater renewable integration in both the main and the isolated grids. The Bill also proposed an attractive FIT rate for on-grid and off-grid investors.[29] It was enacted into law in 2013 for the sole purpose of promotion, development, sustainable management, and utilization of renewable energy sources. The Ministry for Energy recommends national targets for the use of renewable energy resources as set out in the Act.

The Renewable Energy Association of the Gambia (REAGAM) promotes collaboration between several enterprises to promote renewable energy projects, including micro-hydroprojects. However, Gambia's future development of her hydropower resources is promising when considered against the background of regional cooperation. The country, although lacking hydropower potential, is developing the Gambia River Basin Development Organization (OMVG) in collaboration with the governments of Senegal, Guinea-Bissau, and Guinea to construct two large-scale hydropower generation units at Sambagalo (Senegal) and Kaleta (Guinea).[30] The Electricity Strategy and Action Plan identified lack of regional interconnection as a major challenge for hydropower development, but noted that the Gambia could play a significant role in achieving the regional targets for increased renewable energy in the overall electricity mix under ECOWAS to increase the share of renewable energy in the overall electricity mix, including large hydro. Nevertheless, it acknowledged Gambia is in a constrained position due to a small and unstable grid that places a cap on the

---

29. Section 13.
30. IRENA Gambia energy readiness assessment.

amount of renewable energy that can be incorporated into the grid. This necessitates the need for the Gambia to take advantage of regional interconnection and revisit renewable energy targets against the background of such initiative.

Gambia's national environmental management Act was enacted in 1994 to establish the National Environment Management Council. The Act mandated the Council to coordinate all policies in both public and private sectors that are likely to have significant impact on the environment and promote the integration of environmental considerations into all aspects of social and economic planning.[31] The Council was also mandated to harmonize the plans and policies of the various sectors dealing with the environment and approve all environmental plans and policies, and promote the use of renewable sources of energy. Even though the Act came a little later than the UNFCCC in 1992, the Renewable Energy Bill of 2012 was enacted into law in 2013 to establish legal, economic, and institutional basis to promote the use of renewable energy resources as envisaged by the environmental Act. Gambia's Climate Change and Development Plan (Jaiteh and Sarr, 2011) noted that climate change and variability is expected to have a direct effect on energy supply and demand, energy-related infrastructure, and a projected increase in the demand for electricity.

4.6 Cape Verde

Cape Verde lacks fossil fuel resources, but reliable renewable energy sources are still largely unexploited. The country's energy policy objective is set out in the National Energy Plan 2003–2012 as the guarantee of satisfaction of the energy need and at a cost that contributes to improve the well being and the quality of life of the population, and for the competitiveness of the national economy. This objective was restated in the Growth and Poverty Reduction Strategy Paper issued in 2005.

In 2010, a Renewable Energy Plan (2010–2020) was developed with the support of Gesto Energia S.A., an international company focused on energy consulting and renewable energy project development. The Plan aimed to generate 50% of the country's energy from renewable resources by 2020. This target requires the committed involvement of the private sector, but there is no framework in place for a feed-in-tariff mechanism. A framework for Independent Power production was put in place in 2011 as set out in Law n1/2011. The Law created tax exemptions on imported renewable energy equipment, guaranteed a 15-year Power Purchase Agreement (PPA) as part of the framework for microgeneration from renewable energy.

Cape Verde's electricity industry is vertically integrated and generation and distribution of electricity services in most parts of the country is the responsibility of the government-owned utility Empresa de Electricidad e Agua. The General Directorate of Industry and Energy (Direcção Geral da

---

31. Section 7.

Indústria e Energia, DGIE) in the Ministry of Tourism, Industry and Energy, and the multisector Agency for Economic Regulation (Agência de Regulação Económica, ARE) have joint responsibility for management of the energy sector. ARE is an independent regulator and also enforces tariff regulations in the sector.

Prominent among the steps that the government has taken to incentivize renewable energy development is the creation of tax exemptions for imported renewable energy equipment. These exemptions were created by Law n° 20/VII/2007, although most of the provisions of the Law are biased in favor of solar energy and wind generators. This, coupled with the government's commitment to ensuring a viable framework for intellectual property rights, potentially places renewable energy including hydropower in a position to attract and sustain private investment.

Cape Verde's Second National Environmental Action Plan[32] identified the need for adaptation of the institutional and legal framework as a main priority for next few years to ensure a multisector approach to programmes and implementation, and to promote the integration of the environmental concerns in socioeconomic development planning. However, the plan did not contain a clearly defined strategy to combat climate change. The second National Communication on Climate Change of Cape Verde to the UNFCCC (Minister of Environment, Rural Development and Marine Resources, 2010) substantially reiterated the Environmental Action Plan.

Cape Verde adopted the ECOWAS Regional Renewable Energy Policy and Energy Efficiency White Paper in 2012, and hosts the ECOWAS Centre for Renewable Energy and Energy Efficiency.

4.7 Mali

The government of Mali set a national policy goal to drive the country's sustainable development through the provision of affordable energy services in order to increase access to electricity and ensure better socioeconomic outcomes. The National Energy Policy (Politique Énergétique Nationale (PEN)) was issued in 2006 to contribute to this goal, and a National Strategy for the Development of Renewable Energy was also adopted in the same year as a 10-year plan to support the widespread use of renewable energy technologies and equipment to increase the share of renewable energies in national electricity generation up to 10% by 2015.

Mali is one of the six pilot countries worldwide and one of the three in Africa to benefit from the Scaling-Up Renewable Energy Program in Low-Income Countries (SREP), financed by the Climate Investment Fund. SREP was designed to demonstrate the economic, social, and environmental viability

---

32. Ministry of Environment and Agriculture, Second National Environmental Action Plan 2004–2014, Cape Verde.

of low-carbon development pathways in the energy sector in low-income countries.[33] The projects proposed by the plan include expansion of Mali's solar photovoltaic, minihydro, and biofuel technologies, and the sustainable transformation of Mali's renewable energy sector (Ministry of Energy and Water Resources National Department of Energy, 2011).

The Strategic Plan for reduction of Poverty for 2007–2011 (Cadre Stratégique de Croissance et de Réduction de la Pauvreté) embodies an inclusive strategy poverty reduction through the provision of a cohesive structure for development partners to align their interventions. The Environment and Sustainable Development Agency (AEDD) is responsible for coordinating Mali's national climate change responses. The national climate change policy was certified by the Council of Ministers in October 2014, and aspects of the climate change policy and strategy have been integrated into the 2012–17 Strategic Framework for Growth and Poverty Reduction. Mali has a Climate Fund that was set up in partnership with UNDP, and it received its first contribution from Sweden in 2013 but became functional in 2014.

Mali's National Energy Directorate is responsible for the formulation of energy policy, general planning, and coordination of the activities of sector stakeholders. Power generation, transmission, and distribution are the responsibility of Énergie du Mali (EDM-SA), while the Electricity and Water Regulatory Commission responsible for pricing, consumer protection, and regulatory compliance. The Agency was created in 2000 and is under the supervision of the Prime Minister's Office. The regulatory barriers to Mali's renewable energy development, which includes the development of the country's hydropower potential, have been identified as weak planning processes and incomplete framework for public-private partnerships, in particular for utility scale independent power projects. This also includes high take-off cost of RE projects, limited human resource capacity, and limited studies into RE sources for electricity generation.

The constitution of Mali[34] guarantees the right to a healthy environment. "Programme d'Action National d'Adaptation aux Changements climatiques," adopted in 2007, aims to mitigate the adverse effects of climate variability and change on the most vulnerable segments of the population for a more sustainable development, including renewable energy projects. A national agency for environmental and sustainable development was established in 2010 to ensure the implementation of the environmental policy, overseeing the integration of environmental aspects across sectors. Despite Mali's strides in the area of climate change policy, there is no clear direction as to how climate change measures and hydropower policy can overlap to guarantee more holistic outcomes for the country.

---

33. http://www.climatefundsupdate.org/listing/scaling-up-renewable-energy-program (Retrieved 9 June 2016).
34. Constitution of Mali 1992.

4.8 Senegal

In 1998, Senegal's Electricity Law (98–29) was promulgated to provide the sector's legal, regulatory, and institutional framework as part of the government's reforms to promote private sector participation in the electricity supply industry. The law also promotes private sector involvement in electricity generation and distribution through concessions and licenses. Senegal developed its National Adaptation Programme of Action (NAPA) in 2006, following the recommendation to least-developed countries to create policy frameworks that enable them to communicate more clearly about their vulnerabilities and priorities for adaptation (Lo and Tumusiime, 2013). A Renewable Energy Law was enacted as a State law in 2010.[35] Its main objective was majorly to create incentive schemes for renewable energy, including the creation of a legal framework for the development of renewables and promotion of all the means of production, storage, distribution, and consumption for domestic and industrial needs in urban areas as in rural areas.[36] A variety of fiscal tax reliefs are embedded within the law to scale up private sector participation and investment in renewables. Total tax exemption is available for the purchase of renewable energy materials and equipment intended for domestic use. This also includes total tax exemption for the purchase of materials, equipment, and for exploitation and/or research in renewable energy.[37]

A major objective of the law was also to establish Senegal's feed-in tariff scheme by establishing a favorable incentive framework for the purchase and remuneration of electricity produced from renewable energy sources.[38] The Act defines renewable energies to include hydroelectric energy from underwater currents, tidal energy from the movement of water created by the tides (variations in sea level, tidal currents), and small hydro from the conversion of a waterfall or the current of a river.[39] While the Law established the conditions of purchase, sale, and payment for electricity produced from remuneration of renewable energies, it left the power ranges of renewable energy types that are within the scope of this Act to be defined by Decree.[40] Two implementing decrees were also issued for the Renewable Energy Law: Decree No. 2011–2013 and Decree No. 2011–2014. The former provides the framework for PPAs and FITs, while the latter provides the tariff framework for renewable electricity.

Major decisions relating to the energy sector, particularly on-grid electricity, are taken by the Council of Ministers, while the decisions related to renewable electricity are taken by an Interministerial Committee on Renewable Energy (Comité Interministériel sur les Energies Renouvelables). This is made up of

---

35. Law No. 201021 of 20 December 2010 Concerning the Outline Law on Renewable Energies.
36. Article 3.
37. Article 8.
38. Article 3 and 14.
39. Article 1.
40. Article 9.

the key actors including the Ministry of Energy (MoE), Ministry of Renewable Energy (MER), CRSE, ASER, and Société Nationale d'Electricité du Sénégal—SENELEC. The Ministry of Renewable Energy has responsibility for renewable electricity policy development.

The national electricity utility (Société Nationale d'Electricité du Sénégal—SENELEC) is a state-owned enterprise that has a monopoly for transmission and distribution. It also owns about half of the generation capacity, with the remainder being owned by Independent Power Producers (IPPs), which generate electricity and sell it exclusively to SENELEC. The Agency for Rural Electrification was established in 2000 to regulate off-grid rural electrification and implementation of government's plans for rural electrification, which is a 30% rural electrification rate by 2015, and a 60% rate by 2022 reach a 30% rural electrification rate by 2015, and a 60% rate by 2022.[41]

Senegal's commitment to renewable energy is also evident in the establishment of a dedicated Ministry for Renewable Energy and interagency cooperation is made possible through the establishment of an Interministerial Committee on Renewable Energy (CIER).

Implementation of the Senegal's National Adaptation Programme of Action is led by the National Committee on Climate Change under the Ministry of the Environment and the Protection of Nature. Like most of the other ECOWAS countries, it does not have a comprehensive plan that outlines the challenges as well as integrated responses to climate change as a multisector problem.

5. CONCLUSION AND RECOMMENDATIONS

Hydropower in renewable energy is not exclusive, but where a policy is not targeted exclusively at only one kind of renewable resource, it may be inferred from the ordinary definition of renewable energy that it includes small hydro systems. Hydropower policy is largely embedded in a national energy policy together with government's goals and objectives for other forms of energy. However, hydropower policy is classified with renewable energy. As a result, the stated targets for hydropower are also contained in larger renewable energy targets, broader regulatory framework for energy, and are often not sufficient to drive development of smaller hydropower systems. In many ECOWAS countries, there are large and well-established hydropower projects that date back to the earliest years of electricity generation in those countries. These projects were the basis for government monopoly in the sector and the subject of electricity sector reform that swept across the region. Further, there are no specialized or exclusive agencies for the promotion and regulation of hydropower. Climate Change and environmental protection as well as hydroelectric

---

41. Renewable Energy & Energy Efficiency Partnership, Senegal 2014. http://www.reeep.org/senegal-2014 (Retrieved 6 July 2015).

regulation are provided for and regulated by different institutions. Cohesion in policy and cooperation among institutions can strengthen mitigation and adaptation measures.

Legislation can address many of the barriers to the use and development of hydropower, and as such is a strong panacea for many of these barriers—from creating regulatory certainty, appropriate incentives to investment-friendly financial mechanisms. Since hydropower use and development cannot be divorced from larger energy policy nor can hydropower be separated from the energy mix because it forms a critical part of it, advancing its use and development can be addressed by overall energy policy. Hydropower use and development can be used to achieve sustainable gains in the environment.

Most countries in the region have implemented energy sector reforms mostly to promote private investment in the sector. However, continuous effort is needed to design and implement measures targeted at incentivizing the sector to grow and sustain the private sector. It is discernible from the study that countries place premium on universal access to modern energy services. Therefore, if the policies and regulatory mechanisms are designed or sustained to enhance the continued development of hydropower, the chances of realizing universal access are high for the regions as hydropower can contribute to this goal. All the states examined have energy policies or laws that were issued in the 2000s and most adopt the universal access to energy service justification.

All the policy goals, strategic plans, and objectives have 2020–30 deadlines, indicating that actions for hydropower development must be in line with these the multisectoral plans toward the deadline goals.

Water, energy, and climate change are inextricably linked. Enforcement of already-existing environmental legislations especially environmental impact assessment legislation will ensure that new hydropower projects satisfy the relevant standards to ensure minimum negative impact. There is need for hydro-specific policy or legislation to ensure aggressive use and development. Synergizing energy and environmental policy to achieve the environmental sustainability necessary for continuous hydropower development.

Access to electricity in rural areas can be achieved through development and use of smaller hydropower systems for off-grid energy systems. Cumulative national actions can shoot the region far in universal electricity access. In the same vein, no country can stand alone if the region is to advance in a unanimously agreed direction as evident in the treaties. Furthermore, national commitments are necessary to achieve sustainable outcomes with regional or international arrangements. At present, there is no dedicated regional agreement or institution dedicated to climate change adaptation and mitigation strategies coordination among the Member States.

ECOWAS does not have a dedicated institution for policy coordination and monitoring or strategic information dissemination. Given that the impacts of climate change can be multisectoral and complex, the policy strategies must also take the same dimension. All ECOWAS States have ratified the NFCCC and the Kyoto Protocol, but only Nigeria has a climate change Bill. The Bill was drafted in 2008, but is yet to be enacted into law. Overall, countries will need to design and adopt economic development pathways that are less dependent on fossil fuels, because proactive planning reduces the impacts of climate change. It is in this regard that hydropower presents a most suitable alternative.

Climate change is a complex multisector phenomenon, and hydropower intersects energy, environment, climate change, trade, and development. So must policies, legislation, and institutions to regulate its use and development. Regional and international commitments matter, but national efforts should take priority. There is need to mainstream hydropower, climate change, and development policies as a response to the current disjointed or mutually exclusive approaches to all these areas places the region in a disadvantaged position; first to achieve national priorities and to leverage on international and regional opportunities. Most countries examined are still at the stage of setting policies in place.

Through their various commitments, ECOWAS States have taken the first steps toward mainstreaming climate change into environment and energy policy and addressing climate change as the complex challenge that it is. Beyond this, they will require a comprehensive approach to climate change as a multisector challenge in addition to consistent action to ensure that the present structures deliver useful outcomes.

REFERENCES

Adeyemo, B. West African Power Pool, Being a Presentation Made at the South Asia Regional Workshop on Competitive Electricity Markets, Colombo, Sri Lanka, March 18, 2014.

Dernbach, J.C., Kakade, S., 2008. Climate change law: an introduction. Energy Law J. 29 (1), 1–31.

Economic Community of West African States, 1975. Treaty of the Economic Community of West African States. http://www.comm.ecowas.int/sec/?id=treaty&lang=en (Retrieved 19 July 2015).

Environmental Protection Agency of Liberia, 2008. National Adaptation Program of Action.

Environmental Protection Agency of Liberia, 2013. Initial National Communication, Monrovia, Liberia.

Government of Ghana, 2011, Ghana's Second National Communication to the UNFCCC. Accra: Environmental Protection Agency.

Government of Sierra Leone Ministry of Transport and Aviation, 2007. National Adaptation Programme of Action (NAPA).

Jaiteh, M.S., Sarr, B., 2011. Climate change and development in the Gambia: challenges to ecosystem goods and services.

Lo, H.M., Tumusiime, E., 2013. Oxfam America Research backgrounder the influence of US development assistance on local adaptive capacity to climate change insights from Senegal.

Minister of Environment, Rural Development and Marine Resources, 2010. Second national communication on climate change of Cape Verde.

Ministry of Energy and Water Resources National Department of Energy, 2011. Renewable energy Mali: achievements, challenges and opportunities.

Ministry of Environment of the Federal Republic of Nigeria, 2003. Nigeria's first national communication under the United Nations framework convention on climate change.

United Nations, 1972. Stockholm declaration. www.Unep.Org/documents/default.Asp? DocumentID=97 (Retrieved 10 July 2016).

World Commission on Environment and Development, 1987. Our Common Future. Oxford University Press, Melbourne.

Chapter 12

Re-engineering Hydropower Plant for Improved Performance

Eric A. Ofosu*, Mark Amo-Boateng*, Martin K. Domfeh*, Robert Andoh[†]
*University of Energy and Natural Resources, Sunyani, Ghana, [†]AWD Consult. Inc., Portland, ME, United States

Chapter Outline

1. Introduction	189	4. Optimizing Intake Configurations to Facilitate Power Generation at Low Water Levels at Akosombo Dam	193
2. The Case Study of the Akosombo Dam	190		
3. Reoperation and Reoptimization of the Akosombo Dam	191	5. Turbine Plant Intake Optimization	193
		References	194

1. INTRODUCTION

Energy is considered as an indispensable requirement of life and for national development. Every day, the world population is rising at an exponential rate and as a result the demand for energy is increasing significantly. The global call for a sustainable environment has made renewable energy resources indispensable. Concerns about the impact of energy resources on the environment, emission of greenhouse gases and its attendant climate change impacts have become a reigning subject in the field of scientific research.

Considering global electricity generation, hydropower accounts for 16% while other renewables account for 2.6% of global annual generation (International Energy Agency, 2016). Hydropower is one of the cleanest sources of energy with untold advantages and disadvantages to mankind and the environment. Despite the enormous benefits from hydropower plants, Hydropower development has come at a huge cost to livelihoods and the environment (Geker, 1999). The construction of dams significantly changes the morphology of the rivers and also the livelihood benefits that inhabitants living around the river gained. The ecosystems of several rivers have been transformed by the construction of hydropower dams, coming at a huge expense to the environment and livelihoods (Tsikata, 2006).

Sustainable Hydropower in West Africa. https://doi.org/10.1016/B978-0-12-813016-2.00012-5
189

It has therefore become necessary to re-engineer the development of hydropower systems, such that the operations of the power systems will mimic the natural flow of the river bodies that have been altered to maintain environmental sustainability and livelihood supports.

Secondly, designers and plant managers are usually confronted with the challenge of vortices at the intake of power plants especially at low water levels. These vortices manifest as swirling fluid flow often with air entrainment. Their development is caused by a number of causative factors: low water levels, sudden changes in direction of flow, sudden steep velocity gradients, and high flow velocities (Yang et al., 2014; Sarkardeh et al., 2010; ASCE, 1995). Generally, vortices have a complex mode of development with several dependent factors, which necessitates the need for CFD and laboratory experiments to unravel the mysteries concerning their occurrences. In order to prevent the formation of vortices, the water level in dam must be kept above the critical submergence (Amiri et al., 2011; Wang et al., 2010). The need to address vortex formations at power intakes stems from the detrimental havocs associated with them and these include turbine vibration and cavitation, energy losses, corrosive effects, and debris entrainment (Andoh et al., 2008; Knauss, 1987).

In ensuring a stable flow to intake with reduced occurrence of vortices, antivortex devices are employed. They come in various types and operate by suppressing the formation of vortices. Their application is however site specific, often requiring experimental and numerical assessment of the problem before selection of the most appropriate antivortex device (Roshan et al., 2009; Knauss, 1987).

This chapter addresses these two important factors that call for the re-engineering of hydropower dams for optimal performance. The Akosombo Dam, which is one of the oldest dams in Africa, is used as a case study for the re-engineering of hydropower plants. The following sections throw more highlights on the two studies.

2. THE CASE STUDY OF THE AKOSOMBO DAM

The Akosombo Dam was constructed in 1965 and is situated at the downstream portion of the Volta Basin and has an installed capacity of 1020 MW (Fig. 1). Table 1 highlights other relevant technical details of the dam.

Two separate studies aimed at re-engineering the Akosombo Dam have been ongoing in the past few years with one fully completed, while the other is just commencing. The completed study looked at the reoperation and re-optimization of the Akosombo Dam to restore downstream ecosystem and improve the livelihoods of the inhabitants. The second study seeks to investigate the development of vortices at low water levels and how this could be addressed using an appropriate antivortex device using the Akosombo Dam as a case study.

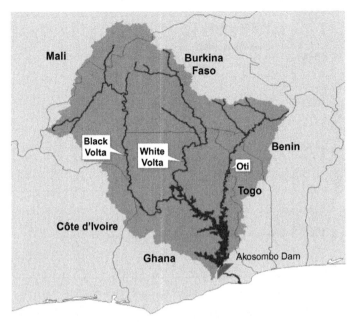

FIG. 1 Location of the Akosombo Dam.

TABLE 1 Technical Details of Akosombo Dam

Maximum height from bedrock	114 m
Area of lake	8502 m^2
Volume of lake	148 × 10^9 m^3
Length of lake	400 km
Max. operating level	84.73 m
Min. operating level	73.15 m

3. REOPERATION AND REOPTIMIZATION OF THE AKOSOMBO DAM

The Akosombo Dam is a storage dam that stores water during seasons of high flows and utilizes it for power generation during seasons of low flows. By virtue of this operation rule, the Akosombo Dam alters the natural regime of the River by storing and releasing water based on electricity demand patterns (Fig. 2). This affects the natural dynamic interaction between the river and the ecosystem (floodplains, wetlands, lagoons, deltas, estuaries,

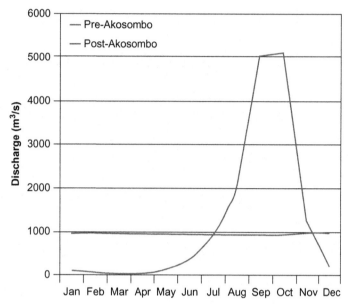

FIG. 2 Pre- and postdam flow regime in the Lower Volta. *(Reproduced with permission from Mul, M., Sidibé, Y., Annor, F., Ofosu, E., Boateng-Gyimah, M., Ampomah, B., Addo, C., 2015. Balancing hydro generation with sustainable ecosystem management. In: Water Storage and Hydropower Development for Africa. Palais des Congrès de la Palmeraie, Marrakesh, pp. 46–50.)*

beach environments, and mangroves) prior to the dam construction. Prior to the construction of the Akosombo Dam there was a vibrant fish industry, lucrative shell fishery dominated by women, active floodplain agriculture, which depended on rich alluvial deposits and control of weeds from the salinity regime during low flows. Presently, there is a drastic reduction in recession agriculture and clam fishing, explosion of exotic weeds that have destroyed the fishing industry, loss of many commercially viable fish species, increased Bilharzia causing snail vectors, the formation of permanent sandbars at the Estuary due to lack of high flows, which seasonally breaks the sandbars, and increase in coastal erosion because of lack of sediment deposition at the estuary, which are used to replenish the coastal erosion (Tsikata, 2006).

The dam reoperation concept is described in Fig. 3. In this concept, the dam is operated in such a way that water released for Hydropower Generation will create a downstream condition that tries to mimic the predam conditions as shown in Fig. 2, thereby restoring the lost ecosystem and livelihoods.

The study looked at the technical, environmental, economic, and socioeconomic feasibility of reoperating the Akosombo Dam. The reoperation was found to be technically feasible with constraints on system stability, economically unfavorable and socioeconomically challenging but having positive environmental impact (Ofosu et al., 2017).

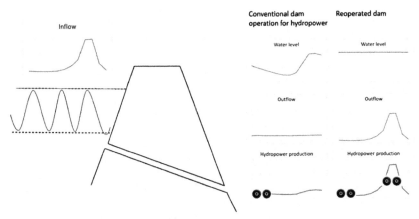

FIG. 3   Dam reoperation concept. *(Reproduced with permission from Ofosu, E.A., Mul, M., Boateng-Gyimah, M., Annor, F., Ampomah, B.Y., 2017. Overview of reoperation and re-optimisation of Akosombo and Kpong Dams Project. In: Ntiamoa-Baidoo, Y., Ampomah, B., Ofosu, E.A. (Eds.), Dams Development and Downstream Communities: Implications for Re-Optimising the Operation of the Volta Dams in Ghana. Digibooks Ghana. pp. 3–24.)*

4. OPTIMIZING INTAKE CONFIGURATIONS TO FACILITATE POWER GENERATION AT LOW WATER LEVELS AT AKOSOMBO DAM

Secondly, a peculiar challenge that has bedevilled the Akosombo Hydropower facility for the past two decades has been the issue of perennial low water levels (as evidenced in Fig. 4) often attributed to climate change, seasonal drought conditions, and upstream catchment developments. This situation has often plunged the nation into recurrent power crises (Gyamfi et al., 2015; Fiagbe and Obeng, 2006; Mul et al., 2015; Gyau-Boakye, 2001). An ongoing project being championed by University of Energy and Natural Resources, Sunyani-Ghana, with support from University of Exeter, United Kingdom, seeks to investigate the development of vortices at low water levels and how this could be addressed using an appropriate antivortex device.

5. TURBINE PLANT INTAKE OPTIMIZATION

Much of the studies in predicting vortices at intakes have utilized commercial CFD tools, but this ongoing study seeks to use Open-source OpenFOAM CFD tool to elucidate the developments and mitigation of vortices at the intake of Akosombo Dam. The research study involves modeling of the problem case in the OpenFOAM CFD tool and the corresponding validation using a laboratory-scale prototype. A range of antivortex devices will also be assessed experimentally in order to select the optimal antivortex device. The study will be concluded with the assessment of the impact of the selected antivortex device on energy generation and a further benefit-cost assessment of the optimal antivortex device.

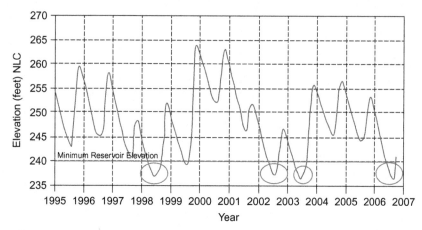

FIG. 4 Akosombo Dam reservoir level variation (Brew-Hammond and Kemausuor, 2007).

REFERENCES

American Society of Civil Engineer (ASCE), 1995. Guidelines for Design of Intakes for Hydroelectric Plants. Committee on Hydropower Intakes of the Energy Division.

Amiri, S.M., Amir, R.Z., Roshan, R., Sarkardeh, H., 2011. Surface vortex prevention at power intakes by horizontal plates. Proc. Inst. Civ. Eng. Water Manage. 164 (4), 193–200.

Andoh, R., Osei, K., Fink, J., Faram, M., 2008. In: Novel drop shaft system for conveying and controlling flows from high level sewers into deep tunnels. World Environenmental Water Resources Congress, Honolulu, Hawaii. pp. 1–9. https://doi.org/10.1061/40976(316)251.

Brew-Hammond, A., Kemausuor, F., 2007. In: Energy crisis in Ghana: drought, technology or policy? Proceedings from seminar on Energy Crisis in Ghana, Kumasi-Ghana. p. 1.

Fiagbe, A.K.Y., Obeng, D.M., 2006. Optimum operations of hydropower systems in Ghana when Akosombo Dam level is below minimum design level. J. Sci. Technol. 26, 2.

Geker, J., 1999. The effects of the Volta Dam on the People of the Lower Volta. In: The Sustainable Integrated Development of the Volta Basin in Ghana. VBRP, Legon, pp. 119–121.

Gyamfi, S., Modjinou, M., Djordjevic, S., 2015. Improving electricity supply security in Ghana - The potential of renewable energy. Renew. Sust. Energ. Rev. 43, 1035–1045.

Gyau-Boakye, P., 2001. Environmental impacts of the Akosombo Dam and effects of climate change on the lake levels. Environ. Dev. Sustain. 3, 17–29.

International Energy Agency, 2016. http://www.IEA.org. [Online] [Cited: May 29, 2017].

Knauss, J., 1987. Swirling flow problems at intakes. In: IAHR Hydraulic Structures Design Manual 1. Balkema, Leiden.

Mul, M., Sidibé, Y., Annor, F., Ofosu, E., Boateng-Gyimah, M., Ampomah, B., Addo, C., 2015. Balancing hydro generation with sustainable ecosystem management. In: Water Storage and Hydropower Development for Africa. Palais des Congrès de la Palmeraie, Marrakesh, pp. 46–50.

Ofosu, E.A., Mul, M., Boateng-Gyimah, M., Annor, F., Ampomah, B.Y., 2017. Overview of Re-operation and Re-optimisation of Akosombo and Kpong Dams Project. In: Ntiamoa Baidoo, Y., Ampomah, B., Ofosu, E.A. (Eds.), Dams Development and Downstream Communities: Implications for Re-optimising the Operation of the Volta Dams in Ghana. Digibooks Ghana, pp. 3–24.

Roshan, R., Sarkardeh, H., Zarrati, A.R., 2009. Vortex study on a hydraulic model of Godare-Landar dam and hydropower plant. WIT Trans. Eng. Sci. 63, 217–225.

Sarkardeh, H., Zarrati, A.R., Roshan, R., 2010. Effect of intake head wall and trash rack on vortices. J. Hydraul. Res. 48 (1), 108–112.

Tsikata, D., 2006. Living in the Shadow of the Large Dams, Long Term Responses of Downstream and Lakeside Communities of Ghana's Volta River Project. African Social Studies Series. Brill, Leiden/Boston.

Wang, Y., Jiang, C., Liang, D., 2010. Investigation of air-core vortex at hydraulic intakes. J. Hydrodyn. 22 (5), 696–701. https://doi.org/10.1016/S1001-6058(10)60017-0.

Yang, J., Liu, T., Bottacin-Busolin, A., Lin, C., 2014. Effects of intake-entrance profiles on free-surface vortices. J. Hydraul. Res. 52 (4), 523–531. https://doi.org/10.1080/00221686.2014.905504.

Chapter 13

Hydropower Generation in West Africa—The Working Solution Manual

Amos T. Kabo-bah*, Chukwuemeka J. Diji†, Kofi A. Yeboah*

*University of Energy and Natural Resources, Sunyani, Ghana, †University of Ibadan, Nigeria

Chapter Outline

1. Introduction	**197**	3.3 Operation and
1.1 Importance of Hydropower		Maintenance—Use of
in the World, Africa, and		Robotics and Digital
Ghana	197	Innovations, etc 203
2. Challenges	**200**	3.4 Life-Cycle Integration
2.1 Climate	200	With Other Sources
2.2 Infrastructure	201	of Energy 203
2.3 Operation and		3.5 Water-Energy Nexus 204
Maintenance (O&M)	201	3.6 Extreme Event—Preventing
2.4 Life Cycles	201	failure 204
2.5 Extreme Event-Dam failure	201	3.7 Other Considerations 205
3. Proposed Working Manual	**202**	**4. Conclusion** **205**
3.1 Climate	202	**References** **206**
3.2 Infrastructure	202	

1. INTRODUCTION

1.1 Importance of Hydropower in the World, Africa, and Ghana

Globally, hydropower has become an important option for power supply for a number of reasons. The International Finance Corporation (IFC, 2015) enumerates the following reasons. First, it is a renewable energy resource that can contribute to sustainable development by generating local, inexpensive power. Second, hydropower reduces reliance on imported fuels that carry the risks of price volatility, supply uncertainty, and foreign currency requirements. Additionally, hydrosystems can offer multiple cobenefits including water

Sustainable Hydropower in West Africa. https://doi.org/10.1016/B978-0-12-813016-2.00013-7

197

storage for drinking and irrigation, drought preparedness, flood control protection, aquaculture, and recreational opportunities, among others. Lastly, hydro allows for synergies with other renewable technologies, especially wind and solar to be integrated to the system by providing rapid-response power when intermittent sources are offline, and pumped energy storage when such sources are generating excess power.

Hydropower is an attractive energy source with minimal operational emissions of greenhouse gases. In addition, there are no fuel charges and the civil works have a long useful life. However, large dams may necessitate population displacement and can impact on the ecology of the basin (Whittington and Gundry, 1998; Barber and Ryder, 1993). Notwithstanding the fact that hydropower plants require massive capital investments, IFC (2015) states that they offer extremely low operating costs and long operating lifespans of 40–50 years that can often be extended to 100 years with some rehabilitation. The result is competitive production costs for electricity.

Hydropower generated 16.4% of the world's total electricity supply and is the world's leading source (71%) of renewable electrical energy (World Energy Council, 2016).

The International Hydropower Association (IHA), reported that in 2016, the total global hydropower generation was estimated at 4102 TWh, with an estimated operational hydropower capacity of 31.5 GW, including pump storage, bringing the world's total installed capacity to 1246 GW (IHA, 2017). IFC (2015) reports that Sub-Saharan Africa has the largest energy access deficit, with over 400 GW of undeveloped hydropotential.

In the year 2016 alone, IHA (2017) reports that 3413 MW was added to the hydropower generation capacity of Africa with the most notable additions being the 1870-MW Gibe III project in Ethiopia and the 1332-MW Ingula pumped storage project in South Africa. Most countries in Africa are making significant strides by harnessing their hydropower potential either internally or among nations that share transboundary rivers.

Africa's strategic energy vision, as reaffirmed in the Maputo Declaration in 2010, is to develop efficient, reliable, cost-effective, and environmentally compliant infrastructure for the physical integration of the continent and to enhance access to modern energy services for the majority of the African population. The huge hydropower potential in Africa presents opportunities to use the technology at the regional, national, and local levels, in the form of large, small, micro-, and mini-hydropower systems. The development of hydropower systems in Africa features prominently in the infrastructure development plans of the African Union Commission (AUC), the regional economic communities, as well as the regional power pools (IHA, 2015).

Out of the total of 20.3 GW of hydropower installed in Africa, about 25% is located in West Africa, 23% in North Africa, and the remaining 51% in Central/South/Eastern Africa (Kalitisi, 2003). Until late 1998, most of Ghana's electricity was produced from two hydrodams at Akosombo and Kpong, which had a combined installed capacity of 1072 MW (Kalitisi, 2003), until the addition of the 400-MW Bui hydro dam in 2013. However, Kalitisi (2003) estimates that Ghana may have a resource potential of additional 2000 MW of hydropower. Ghana has potential for exploitable medium hydropower sites with the potential of some sites estimated to be over 75 MW (Kabo-bah et al., 2016). However, data obtained from VRA (2017) suggest that Ghana's hydropower constitutes 43.3% of the total installed capacity for electricity, currently.

One of the key challenges of power supply in West Africa regardless of the type is the low quality of service. The drought in 1998 resulted in low water levels in the Akosombo reservoir that led to the energy crises in Ghana, Benin, and Togo (Gnansounou et al., 2007). This negatively affected the country's economic development during that period. Quite recently also in 2014, energy crises had been reported in Ghana as a result of the inadequacy of the Akosombo and the Bui hydroelectric plants to meet the needed demands in Ghana (Boateng, 2014; Appiah, 2015). In 2001, Nigeria experienced rolling blackouts due to drought and draining of the Kainji, the largest hydropower reservoir in the country. Similarly, Senegal, Guinea, and Mali have also suffered several years from frequent disruptions due to insufficient generation capacities and low reliability of the power plants. Notwithstanding the commissioning of hydropower plants in Manantali in Mali and Garafiri in Guinea, the electricity demand within these countries is still insufficient to meet the current demand and needs. Ivory Coast, on the other hand, is the notable exception with a well adapted to significant extent to its electric supply demand (Gnansounou et al., 2007).

As Gnansounou et al. (2007) stated, countries like Ghana, Nigeria, Benin, and Togo overly depend on hydropower, with an over reliance on multiannual variability in hydrological conditions and these countries are bound to continuously experience deficits unless additional energy power supplies are tapped into. For instance, between 1960s and 1990s, temperatures within the subregion rose by 10°C, the mean annual rainfall and runoff dropped by more than 30% resulting in what Hulme et al. (2001) described as the most dramatic multidecadal climate variability.

The key resource for hydropower generation is runoff, which is dependent on precipitation. The future of global climate is uncertain and thus poses some risk for the hydropower generation sector. The crucial question and challenge then is what will be the impact of climate change on global hydropower generation and what are the resulting regional variations in hydropower generation potential? (Hamududu and Killingtveit, 2012).

Hydropower plays an important role in world energy supply and it is significant in combating global climate change. At present, world hydropower development has entered a new phase and faces challenges from perspectives of environment, water resources, and others. Hydropower can be part of a sustainable energy future if designed and operated in a manner that avoids or minimizes impacts on people and vital river functions (IHA, 2015). This is necessary for stakeholders in the West African subregion, which is mostly affected by harsh climatic impacts, to deal with perceptions of climate risk and hydropower sector's resilience to the impacts of climate change.

2. CHALLENGES

Generally, the challenges in the hydropower sector in West Africa are varied but range from adverse effects of climate change on rainfall-run-off, aging hydropower equipment that have reached or are approaching the end of their expected operational life, operation and maintenance (O&M), difficulty with access to finance, low generation capacity and efficiency, and insufficient energy infrastructure among others. There is therefore a need for global cooperation of all partners and stakeholders, on all levels, to tackle these challenges to find ways to remove the barriers in order to maximize the use of the hydropower resource.

2.1 Climate

Africa is one of the continents most at risk from climate change. Observed temperatures have indicated a warming trend since the 1960s, and IPCC projections suggest that annual temperatures in the region will rise by 3.4°C (Ramsay et al., 2014).

It is a well-known fact that the impacts of climate change pose a risk to the realization of the SDGs. Its impacts will intensify the increasing competing water uses for irrigation, municipal water supply, and hydropower to a rapidly increasing global population, of which West Africa is no exception. Most countries in West Africa will become vulnerable to the extreme weather events that may occur from climate change. As a result, these countries will be impacted heavily due to least developed infrastructures that exist.

IHA (2017) findings indicate that most project developers and financing institutions currently lack the capacity to incorporate climate resilience measures into project design and appraisal.

The direct impacts of climate change on hydropower are caused by changes in hydrology and may include reduced or increased annual flows, increased flood runoff (with associated erosion and sedimentation), longer dry seasons, and increased variability or uncertainty (Ramsay et al., 2014).

According to Eberhard et al. (2011), unpredictable weather patterns as a result of climate change reduce hydropower's reliability, thereby increasing the cost of generating and delivering power in Africa. Besada and Sewankambo (2009) state that droughts that occurred in many African countries had been linked to climate change and it adversely affected both energy security and economic growth across the continent. The uncertainties associated with hydrology may impact power generation and revenues (IFC, 2015). Besada and Sewankambo (2009) further asserted that the impacts of drought on the energy sector due to the impacts of climate change were felt primarily through losses in hydropower potential for electricity generation. Furthermore, Cole et al. (2014) hint that results from climate modeling predictions puts planned projects in most African countries in the balance due to viability concerns.

2.2   Infrastructure

As stated by IFC (2015), the available flow, head, and project location deter-mines the type of infrastructure, machinery to deploy and its associated costs, during the engineering, procurement, and construction stage. Moreover, the cost of developing some potential site considering the project location, geology, available head and flow as well as the existing power market tariffs does not make it an economically viable venture. Furthermore, there exists some level of energy loss with the transmission of power in the subregion, which leads to low productivity.

2.3   Operation and Maintenance (O&M)

As stated by IFC (2015), O&M refers to all of the activities needed to run a hydropower plant (HPP) scheme, except for construction of new facilities. Maintenance costs represent a greater chunk of total operational costs. IFC (2015) estimates annual O&M costs to range from 1% to 4% of investment costs. It however states that these costs are exclusive of major electromechani-cal equipment replacement, which would be required a couple of times during the lifespan of the hydropower plant (HPP) and would therefore raise average O&M costs to US$45/kW/year for large HPP; US$52/kW/year for small HPP. With aging HPP facilities, O&M costs rise over time.

It is therefore very obvious that capital is needed for an effective O&M to maximize the benefits of a hydropower and extend its life cycle.

2.4   Life Cycles

As stated earlier by IFC (2015), the life cycle of most hydropower plants is be-tween 40 and 50 years and can be extended to 100 years with some retrofitting. With most global hydropower infrastructure expected to reach the end of their life cycle by 2050 (IHA, 2017), stakeholders in the hydropower sector in West Africa are saddled with the challenge of choosing the appropriate rehabilitation and renewable integration methods that will expand the productive life of the existing facilities.

2.5   Extreme Event-Dam failure

The possibility of a dam failure is ever present in the operation of a hydropower facility. Extreme floods that are not appropriately managed, can lead to overtop-ping and subsequent dam failure. Other factors that pose a challenge that may lead to dam failure are extreme seismic events and undetected structural defects in the retaining walls of reservoirs. Seismic-related failures may occur if the appropriate factor of safety to cater for seismic activity is not carefully considered in the design and construction of the HPP. Undetected structural defects may be attributed to poor inspections and preventive maintenance (PM).

3. PROPOSED WORKING MANUAL

There is the need to develop in a sustainable manner as enshrined in the Sustainable Development Goals (SDGs). As stated by IHA (2015), hydropower is relevant to several of these goals, notably goal 7 (ensure access to affordable, reliable, sustainable and modern energy for all), goal 13 (take urgent action to combat climate change and its impacts), and goal 6 (ensure availability and sustainable management of water and sanitation for all). Finding the right balance between sustainably developing hydropower in the West African subregion, and its associated challenges with climate change, infrastructure among others, should be the way forward in the hydropower sector.

3.1 Climate

Reservoirs of hydropower plants can be used as a tool to minimize the impacts of climate change, and in turn play an active part in the achievement of many of the SDGs. Hydropower reservoirs tackle SDG#13 that calls for an urgent action to combat climate change and its impacts. With SDG#7, hydropower provides that affordable, reliable, flexible, and renewable energy the SDG calls for. Hydropower systems also have the added flexibility of augmenting its output by connecting with wind and solar energy, thereby supporting renewable energy systems (IHA, 2017).

Perspectives obtained from climate modeling should also be used to design and modify proposed and existing projects to make them viable. To ensure hydropower projects are resilient in the face of uncertainty of climate change impacts, IHA (2017) has developed a tool, Hydropower Sustainability Assessment Protocol, which provides practical guidelines to achieve climate resilience in the planning, design, construction, and operation of hydro projects. This tool, as stated by IHA (2017), would be very useful in measuring the sustainability of projects across a range of social, environmental, technical, and economic considerations in West Africa (IHA, 2017).

3.2 Infrastructure

The development of infrastructure in the hydropower sector is crucial in providing the bedrock for a nation's development. This is because it offers the cheapest competitive power pricing tariff on the energy market, needed by industries and businesses to make profits and create employments for the teeming population for nations across West Africa. Hydropower infrastructure must be rehabilitated to keep pace with energy demands of nations, and to prevent unreliable power supply and economic downturn. According to Ramsay et al. (2014), planned projects should take reservoir shape into consideration in the design of HPPs in order to reduce evaporation and maximize power potential.

Industry best practices must be chosen in the development of hydropower potential in the subregion, so as to select the best design and appropriate infrastructure necessary. Best practices in infrastructure selection are important considering the huge investments needed in hydropower developments and lack of cash to finance such projects. As spelt out by Hollingworth (2014), integrated national and regional infrastructures for energy supply, efficient transmission, and distribution systems must be used to make our energy systems efficient and sustainable.

3.3 Operation and Maintenance—Use of Robotics and Digital Innovations, etc

IFC (2015) emphasizes that existing and planned hydropower facilities must consider the inclusion of software-based control and monitoring systems to facilitate remote maintenance, diagnostics, trouble shooting, parameter adjustments, etc. Spare parts inventory must be optimized to reduce costs and maintain plant availability and reliability. Scheduled preventive maintenance must be routinely followed to prevent system failures. Practical steps should be taken to invest in a combination of condition-based, reliability-centered preventive maintenance procedures to minimize individual generating unit downtime and supply a level of operational reliability such that grid demands are met with available generation.

The issue of hydropower asset facility management is increasingly becoming important due to the growing number of hydropower assets reaching the end of their expected life. It is expected that the entire fleet of existing hydropower equipment in Africa will require modernization by 2050. HPP facility managers must deploy hydropower asset facility management to enable access to data-based actionable insights, and to increase and maximize the value of hydropower assets. IHA (2017) emphasizes the importance of digital innovations in hydropower operations. Digitization of hydropower plants, process control systems, and ancillary networks will optimize asset management and performance. It is expected that digitizing hydropower plants will facilitate seamless connection with other renewable resources such as solar and windmills to provide increased flexibility and enhanced control for ancillary services, i.e., frequency control, balancing services among others (IHA, 2017).

3.4 Life-Cycle Integration With Other Sources of Energy

Personnel must be trained on proper O&M planning and day-to-day maintenance to ensure long-term economic HPP operation. Properly planned O&M prevents prolong forced downtimes and system collapse. That is the only way to ensure existing and planned HPPs perform effectively to their expected useful life.

Floating photovoltaic (PV) solar panels, also known as "floatovoltaics," can be deployed on storage reservoirs to generate energy, which can be integrated into the hydropower generation grid. Innovative hybrid projects of this nature that integrates renewable technologies, not only offer stable power to the grid but also increase efficiencies, from the stable temperature regimes. Floating PV on the reservoir surface is the most novel example of the multiple services that a reservoir can provide, by producing clean, renewable energy with a higher efficiency than conventional land-based solar PV, while reducing evaporation from the reservoir surface. Additionally, this hybrid solution can be achieved without competition for land acquisition (IHA, 2017).

Finally, the IHA Sustainability Protocol reported by Hollingworth (2014) must be used to assess hydropower projects through the four-project life-cycle stages, i.e., early stage, preparation, implementation, and operation stages, for sustainability.

3.5 Water-Energy Nexus

IHA (2015, 2017), points out that the core of Water-Energy Nexus is hydropower and dams, because it uses water to generate electricity and in some instances, make water available for other needs such as navigation, irrigation, aquaculture, domestic water supply, recreation, and trade spin-offs from commercial activities around, when it is operated as a multipurpose reservoir. Additionally, a multipurpose reservoir can be used as a storage capacity to manage floods, offers storage capacity, provides long-term energy storage during extended droughts, provides better environmental management by trapping excess fertilizer runoff from agricultural lands, and pollution attenuation, particularly in dams where wetlands can be developed.

It is expected that development and role of hydropower will continue to expand in West Africa in the coming years. It is therefore imperative to develop multipurpose reservoirs of that nature to improve the nexus efficiency. Stakeholders in the hydropower sector in West Africa must seek better understanding of the water-energy nexus, especially on how hydropower will efficiently use water and its contributions to manage water scarcity.

Optimizing and sharing the multiple benefits of a multipurpose reservoir will not be achieved unless the decision-making process incorporates all stakeholders from the earliest planning stage and reflects a comprehensive and sustainable approach that integrates the social, environmental, and economic dimensions. Multipurpose reservoirs can be an effective solution to the competing uses over water, land, and energy (IHA, 2017).

3.6 Extreme Event—Preventing failure

The dam and its spillway should be designed to have sufficient discharge capacity for a discharge flow with the requisite statistical return period to prevent overtopping. Scheduled PMs and inspections must be followed religiously as well. The use of new and appropriate diagnostic tools must be considered in conjunction with reliability and condition monitoring logistics available to prevent failure. Controlled spills must be managed with the downstream communities in mind to minimize the negative effects of flooding. Furthermore, there must be use of decision support systems that include life-cycle analysis, environmental impacts assessments, strategic environmental assessment, guidelines and protocols in the planning and operation of all hydropower facilities to prevent extreme events.

3.7 Other Considerations

There is the need to look for ways to increase the implementation of hydropower development. The areas to explore include access to finance, establish effective policies and regulatory regimes to stimulate development and investment in the sector as well as new and better ways to address the environmental and social issues (i.e., displacement of people and effect on biodiversity). Tackling the environmental-social conundrum is very crucial since it may introduce a reputational risk for the developer and financiers as affirmed by IFC (2015). As a subregion, we need to leverage on existing deals and programmes arranged by the African Union Commission such as the Africa-EU Energy Partnership (AEEP) and Programme for Infrastructure Development (PIDA) to improve the hydropower capacity. The AEEP initiative as reported by IHA (2015) is to stimulate the addition of 10,000 MW of new hydropower plants by 2020, taking into consideration both social and environmental standards.

4. CONCLUSION

Many challenges abound in the development and operation of hydropower facilities in West Africa. The most notable being lack of access to finance, viability concerns due to impacts of climate change, O&M issues likely to increase operating cost or reduce expected life as well as the environmental and social issues associated with hydropower projects.

Some working solutions have been suggested to enable industry sector practitioners use as a working manual. These solutions offer ways to mitigate climate change effects on hydropower. The proposed working manual also states the need for an upgrade of aging infrastructure, use of new-age technology in operation and maintenance for better system performance, opportunities for synergies with emerging renewable energy technologies, and ways to combat extreme events. The Regional Commission on Energy must ensure that the proven hydropower potential in most West African nations that are economically viable, is developed into installed capacities. That way, more people will have access to power, which will accelerate industrialization, reduce poverty, and sustain the economies of countries within the subregion.

REFERENCES

Appiah, S., 2015. Prepare for intensified load shedding-GRIDCO. Available from http://www. graphic.com.gh/news/general-news/prepare-for-intensified-load-shedding-gridco.html. (Accessed 18 August 2016).

Barber, M., Ryder, G. (Eds.), Damming the Three Gorges: What Dam Builders Don't Want You to Know. second ed. 1993. Probe International, Earthscan Publications, pp. 1993.

Besada, H., Sewankambo, N.K. (Eds.), 2009. Climate change in Africa: adaptation, mitigation and governance challenges. CIGI special report, Ontario.

Boateng, K.A. (2014): Power Crisis to worsen: Akosombo Level nears minimum. Available from: http://citifmonline.com/2014/08/18/power-crisis-to-worsen-akosombo-level-nears-minimum/ (Accessed 18 August 2016).

Cole, M.A., Elliott, R. and Strobl, E. (2014): Climate Change, Hydro-dependency and the African Dam Boom. Department of Economics Discussion Paper 14-03, University of Birmingham. 46 p.

Eberhard, A., Rosnes, O., Shkaratan, M., Vennemo, H., 2011. In: Forster, V., Briceño-Garmendia, C. (Eds.), Africa's Power Infrastructure: Investment, Integration and Efficiency. IBRD/World Bank, Washington, DC, pp. 352.

Gnansounou, E., Bayem, H., Bednyagin, D., Dong, J., 2007. Strategies for regional integration of electricity supply in West Africa. Energ Policy 35 (8), 4142–4153. https://doi.org/10.1016/j. enpol.2007.02.023.

Hamududu, B., Killingtveit, A., 2012. Assessing Climate Change Impacts on Global Hydropower. Energies 5 (2).

Hollingworth, B., 2014. Training Manual: Comprehensive Options Assessment in Hydropower Development. Network for Sustainable Hydropower Development in the Mekong Countries (NSHD-M). GIZ.

Hulme, M., Doherty R., Nagara, T., New, M. and Lister, D. (2001): African Climate, IAHS Publ., No. 252: p. 307-314

IFC (2015): Hydroelectric power: a guide for developers and investors. Available from: www.ifc. org/.../ifc.../ifc.../hydroelectric_power_a_guide_for_developers_and_investo... (Retrieved 10 March 2017).

IHA (2015). 2015 World Hydropower Congress Handbook. International Hydropower Congress (IHA), London. 55 p.

IHA (2017). 2017 Hydropower Status Report. International Hydropower Congress (IHA), London. 84 p.

Kabo-bah, A.T., Diji, C.J., Nokoe, K., Mulugetta, Y., Obeng-Ofori, D. and Akpoti, K. (2016): Multi-year Rainfall and Temperature Trends in the Volta River basin and their Potential impact on Hydropower Generation in Ghana. Climate 2016, 4, 49; doi: https://doi.org/10.3390/cli4040049. Available from: www.mdpi.com/journal/climate. (Retrieved 5 January 2017).

Kalitisi, E.A.K., 2003. In: Problems and Prospects for Hydropower Development in Africa. The Workshop for African Energy Experts on Operationalizing the NEPAD Energy Initiative. June 2003, Dakar-Senegal.

Ramsay, J., Hollingworth, B., Phousavanh, P., Wironajagud, W., Lan, H.T., Thuon, T., 2014. Training manual: sustaining river basin ecosystems in hydropower development. In: Network for Sustainable Hydropower Development in the Mekong Countries (NSHD-M). GIZ.

VRA (2017): Power generation: facts and figures. Available from: http://www.vra.com/resources/ facts.php (Retrieved 18 April 2017).

Whittington, H.W., Gundry, S.W., 1998. Global climate change and hydroelectric resources. IEE Eng. Sci. Technol. (March)1998.

World Energy Council, 2016. Word Energy Resources 2016, London, United Kingdom. Available at: https://www.worldenergy.org/wp-content/uploads/2016/10/World-Energy-Resources-Full-report-2016.10.03.pdf. (Retrieved 1 March 2017).

Index

Note: Page numbers followed by *f* indicate figures, and *t* indicate tables.

A

Aerosols, 37–38
Africa
  climate change, 34
  energy, 34
  hydropower, 4–5, 4*f*, 34, 197–200
  intra-African collaboration, 31–33
  strategic energy vision, 198
Africa-EU Energy Partnership (AEEP), 205
African Development Bank (AfDB), 42,
  63–64, 68–69
African Union Commission (AUC), 198, 205
Africa Renewable Energy Initiative (AREI),
  38–39
Aga Khan Rural Support Programme
  (AKRSP), 55, 57*t*
Agricultural production, 144–146
Akosombo Dam
  case study, 190
  electric power production, 6
  hydroelectric power, 56, 57*t*, 59,
    99–100, 104
  location, 191*f*
  OpenFOAM CFD tool, 193
  reoperation and reoptimization,
    191–192, 193*f*
  reservoir level variation, 193, 194*f*
  technical details, 191*t*
  turbine plant intake optimization, 193
Alternative energy sources, Ghana, 11–14
Annapurna Conservation Area Project (ACAP),
  54–55, 57*t*
Antidam protests, 56–57, 59
Aquaculture, 110, 115

B

Bator community, 115
Biodiversity, Bui dam on, 114–115
Biomass potential, in West Africa, 85–86
Black Volta, 112–116, 122, 129–130
Bongase, 126, 128–132, 134

BPA. *See* Bui Power Authority (BPA)
Brundtland Commission, 161–162
Brundtland Report, 159–160
Bui Dam
  on biodiversity, 114–115
  fishery
    anticipated impacts on biodiversity,
      114–115
    consumption, 115
    diversity, 116
    efforts, 116
    food security, 115
    impoundment effect, 113–114
    migration, 116
    morphoedaphic index, 111*t*
    people migration, 116
    social and economic impacts, 117
  hydropower, 99–100, 103
  location, 110–113, 112*f*, 123*f*
  resettlement, 125–133, 128*f*
  socioeconomic impacts, 122–124
    changes in livelihood, 127–129
    methodology, 124–125
    number of questionnaires in respective
      villages, 127*f*
    tourism, 133
    waste management, 131–132
    water situation, 129–131
Bui Hydroelectric Power Project,
  55–56, 57*t*, 122
Bui National Park, 123–124, 133–134
Bui Power Authority (BPA), 122–123,
  125–126
  Bongase, 131–132
  goals, 129–130
  insufficient compensation of farmland,
    133–134
  livelihood, changes in, 127–129
  social corporate responsibilities, 130–131
  tourism, 133
Bujagali Dam, 55, 57*t*
Burkina Faso, 96, 99

C

Cape Verde, hydropower use, 180–181
Carbon dioxide (CO_2), 37–38
Carbon financing, 42–43
Carbon markets, and hydropower, 44–45
Carbon Partnership Facility (CPF), 42
Centre for Sustainable Energy Technology, 176
Certified Emission Reduction (CERs), 43–45
CIRCLE model. *See* Climate Impacts Research
    Capacity and Leadership Enhancement
    (CIRCLE) model
CIRCLE Visiting Fellows (CVFs), 29–35
Clean Development Mechanism (CDM),
    38–39, 43–44, 103
Clean energy sources, 43–44
Clean Technology Fund (CTF), 42–43
Climate change, 99–100, 159–160, 200–202
    adaptation, 137–138
    addressing, 164
    in Africa, 34
    in Ghana, 11
    hydropower and, 39–45
    Sub-Saharan Africa, 65–66, 69–71
    and wave energy, 14*f*
    West Africa, 76–77
Climate Change Unit (CCU), 174–175
Climate-driven hydrology, 40–42
Climate Impacts Research Capacity
    and Leadership Enhancement
    (CIRCLE) model
    approach, 30–32
    CIRCLE-funded research, 33–34
    fellowship program, 31–32
    Hydropower Energy Conference, 34–35
    intra-African collaboration, 31–33
    links with the North, 31
    research uptake, 32, 34–35
    Specialist Advisor, 31
    supporting different needs, 30–31
Climate Investment Funds (CIF), 42–43
Climate variability
    adaptation, 137–138, 151–155, 152*t*
    chi-square tests, 150*t*, 153*t*
    local people's perception, 146–151,
        148*f*, 149*t*
    water insecurity, 154–155
Commission's Report, 159–160
Community Water and Sanitation Agency
    (CWSA), 140–142
Computational Fluid Dynamics (CFD), 35
Concentrated solar power (CSP), 81–84
Constraints for adaptation, 137–139, 154–155

Cost of electricity (COE), 19–20
CWSA. *See* Community Water and Sanitation
    Agency (CWSA)

D

Dam failure, 201–202
Department of State for Petroleum, Energy
    and Mineral Resources (DoSPEMR),
    178–179
Designated National Authority (DNA), 173
Discharge of water, 3
Division of Electricity and Renewable Energy,
    176–177
Dumsor, 2–3

E

Early-Career Researchers (ECRs), 30–31
Economic Community of West African States
    (ECOWAS), 161–163, 165–168
Economic feasibility, 19–22
Economic potential, in West Africa, 83
Economics of hydropower plant, 101–103
    base load power case, 104
    benefits, 103–104
Ecotourism, 134
ECOWAS Centre for Renewable Energy and
    Energy Efficiency (ECREEE), 167
ECOWAS Renewable Energy Policy (EREP),
    166–167
Electrical energy supply, in Sub-Saharan
    Africa, 66–69
Electricity Act, 178
Electricity Company of Ghana (ECG), 2–3
Electricity Corporation of Nigeria (ECN), 168
Electricity Feed Law (EFL), 90
Electricity generation, by plant, 9*t*
Electric Power Authority (NEPA), 168
Electric power production, in Ghana, 5–14
Electric Power Sector Reform Act, 172–174
Energy Commission of Nigeria (ECN), 99,
    171–172
Energy consumption, 95–96
Energy security, 88
Energy Services Development (ESD), 55, 57*t*
Energy sources
    in Africa, 34
    diversification of, 78–79
Energy supply sector, 66
Environmental and social (E&S) impacts, 44
Environmental Policy for ECOWAS,
    165–166

Environmental Protection Agency (EPA), 170–171, 176–177
Environment and Sustainable Development Agency, 182

F

Federal Executive Council (FEC), 174
Federal Ministry of Power, 172
Feed-in tariff (FIT), 19–20, 91t
Felou hydroelectric plant, Mali, 102, 105
Financial institutions, in Ghana, 20
Fishers' migration, 116
Fishery, Bui dam
   anticipated impacts on biodiversity, 114–115
   consumption, 115
   diversity, 116
   efforts, 116
   Ghana, 110
   hydropower impacts, 113–117
   impoundment effect, 113–114
   livelihoods, 110, 115, 117
   migration, 116
   morphoedaphic index, 111t
   people migration, 116
   social and economic impacts, 117
Flawed procurement, 56–57
Floatovoltaics, 204
Food security, Bui dam, 115
Fossil fuel, 79–80
Fuel substitution, 47

G

Gambia River Basin Development Organization, 178, 180
Gambia's National Energy Policy, 178
Garafiri project, 199–200
Gas Processing Plant (GPP), 2–3
Geographic conditions, in West Africa, 76–77, 82–83
Geological Survey Department of Ghana, 112–113
Geomorphological feasibility, 18–19
Ghana, 96, 197–200
   Akosombo power plant (*see* Akosombo Dam)
   alternative energy sources, 11–14
   Bui power plant (*see* Bui Dam)
   climate change impacts, 11
   construction costs for 500-MW wave farm, 22t
   cost-benefit analyses, 102–103
   country profile, 5–6
   drainage and partitions of coastline, 7f
   economic feasibility, 19–22
   electricity-generating capacity, 9t, 45–47
   electricity supply in, 2–3
   electric power production, 5–6
   energy demand, 3
   Environmental Protecting Act, 170
   financial institutions in, 20
   fishery sector, 110
   geomorphological feasibility, 18–19
   initial cost, 102, 105
   installed electricity generation capacity, 8t
   legislation, 168–171
   locations of major power plants, 41f
   met-ocean feasibility, 18–19
   Ministry of Energy, 171
   National Climate Change Policy, 170
   Paris Agreement on Climate Change, 165
   renewable energy, 11–14
   total installed electricity capacity as at 2015, 46t
   Wa municipality (*see* Wa municipality)
Ghana Grid Company (GridCo), 6
Global energy consumption, 1
Global Energy Efficiency and Renewable Energy Fund (GEEREF), 42–43
Global Environment Facility Trust Fund (GEFTF), 42–43
Green Climate Fund (GCF), 38–39, 42
Green Energy Revolution Reunion Island (GERRI) Renewable Energy Projects, 55, 57t, 59
Greenhouse gas (GHG) emissions
   hydropower, 37–38, 40, 43–45
   mitigation strategies, 79–80
   Sub-Saharan Africa, 66
Guinea
   Garafiri project, 199
   worm disease, 130

H

Hydel project, 54
Hydraulic energy, 39–40
Hydroelectricity Era, 6
Hydropower development, 96, 189, 197–200
   in Africa, 4–5, 4f, 34
   Akosombo power plant (*see* Akosombo Dam)
   Bujagali Dam, 55
   carbon markets and, 44–45
   case studies, 58–59
   as clean development mechanism tool, 43–44

Hydropower development *(Continued)*
  and climate change, 39–45
  dam failure, 201–202
  factors affecting private sector investment,
    104–105
  financial constraints, 47
  Ghana, 45–47
  holistic view, 54
  impacts on fishery, 113–117
  infrastructure, 201–203
  life cycle, 201, 203–204
  methodology of study, 54
  national frameworks, 168–184
  river flow changes, 5*f*
  Sub-Saharan Africa, 65–66, 69–71
  successful and unsuccessful, 54–57, 57*t*
  sustainable development, 47–49
  sustainable management, 53–54
  technical potential, 97–99, 100*t*
  Tiger Leaping Gorge Dam, 56
  Tipaimukh Dam, 56
Hydropower Energy Conference, 34–35
Hydropower Sustainability Assessment
  Protocol, 170, 202

I

IFC. *See* International Finance
  Corporation (IFC)
Independent Power Producers (IPPs), 184
Infrastructure of hydropower, 201–203
Institutional Strengthening Program, 31–32
Intended Nationally Determined Contributions
  (INDCs), 164–165
Intergovernmental Panel on Climate Change
  (IPCC), 138–139, 160–161
Interministerial Committee on Renewable
  Energy (CIER), 184
International Energy Agency (IEA), 39
International Finance Corporation (IFC),
  197–198, 201
International Hydropower Association (IHA),
  68, 198
International Solar Energy Experts
  Workshop, 15
Intertropical Convergence Zone (ITCZ), 77
Investment cost, 102–103
Involuntary resettlement, 117

J

Jama Host Community, 129–132, 130*f*, 134
Jama Resettlement, 125–126, 130*f*,
  131–132, 132*f*

Johannesburg Plan of Implementation, 79–80
Jubilee Oil Field, 20–22

K

Kumasi Ventilated-Improved Pits (KVIPs),
  125, 131–132, 132*f*, 134
Kyoto Protocol, 43, 162–164, 170, 173,
  185–186

L

Lake Volta, 110
Land use change (LUC) impacts, 81–82
Least developed countries (LDCs), 43–44
Levelized cost of energy (LCOE), 102–103
Liberia Electricity Corporation, 99
Liberia legislation, 176–178
Life cycle, hydropower plants, 201, 203–204
Light crude oil (LCO), 37–38
Lignocellulose, 85–86
Liquefied petroleum gas (LPG), 45
Lower Volta, 192*f*
Low-speed shaft, 16

M

Mali
  Felou hydroelectric plant, 102, 105
  legislation, 181–182
  Manantali, 199
  National Energy Directorate, 182
Manantali, Mali, 199
Maputo Declaration, 198
Marine energy conversion systems, 2, 15–23
  alternative energy sources, 12–13
  challenges, 22–23
  climate change impacts on, 13–14
  economic feasibility, 19–22
  geomorphological feasibility, 18–19
  met-ocean feasibility, 18–19
  prototype, 16–18, 21*f*
Met-ocean feasibility, 18–19
Millennium Development Goals (MDGs),
  79–80, 164–165, 175
Ministry of Energy (MOE), 171, 175, 178–179
Morphoedaphic index (MEI), Bui dam, 111*t*
Municipal Community Water and Sanitation
  Agency, 143

N

Nam Mang 3 Hydropower project, 57*t*
National Adaptation Programme of Action
  (NAPA), 183

National Climate Change Commission for Nigeria, 173
National Committee on Climate Change, 169–170, 184
National Electric Power Policy, 172–173
National Energy Plan, 180
National Energy Policy, 176, 181
National Energy Policy Implementation Strategy, 175
National Environmental Commission of Liberia (NECOLIB), 177
National Environmental Standards and Regulations Enforcement Agency (NESREA) Act, 173–174
National Environment Management Council, 180
National Power Authority (NPA), 175
National Renewable Energy Act, 15
National Renewable Energy Policies (NREPs), 167
National Secretariat for Climate Change (NSCC), 176
Niger, 98
Niger Dams Authority (NDA), 168
Nigeria, 96, 99, 199
　Energy Policy, 171
　First National Communication, 173–174
　legislation, 171–175
　Renewable Energy Master Plan, 172
Nigerian Electricity Regulatory Commission (NERC), 172–174
Nigeria, the Electricity Supply Company (NESCO), 168
Non-Fossil Fuel Obligation (NFFO) law, 90
Nongovernmental organizations (NGOs), 56–57

O

Ocean currents, 16
Ocean waves, 12
OpenFOAM CFD tool, 193
Operation and maintenance (O&M), 201, 203
Organization for Economic Co-operation and Development (OECD), 38

P

Paris Agreement on Climate Change, 165
Photovoltaic (PV) solar power, 83–84
PIDA. *See* Programme for Infrastructure Development (PIDA)
Plan for Renewable Energies and the Rational Use of Energy (PRERURE), 55, 57*t*, 59

Population and Housing Census (PHC), 140
Poverty, 139, 146
Poverty Reduction Strategy Paper, 175
Power-generating companies, privatization of, 22
Power Purchase Agreement (PPA), 180
Power Sector Reform Act, 173–174
Power take-off mechanism, 17*f*
Principal Investigator (PI), 31
Privatization, of power-generating companies, 22
Programme for Accelerated Growth and Development (PAGE), 178
Programme for Infrastructure Development (PIDA), 205
Public Private Partnerships (PPPs), 104–105
Public Utilities Regulatory Authority (PURA), 178–179
Public Utilities' Regulatory Commission (PURC), 171
Public Utility Regulatory Act, 178–179
Pubugou Hydropower Project, 57, 57*t*

R

Renewable energy (RE), 3, 42
　in Africa, 4–5
　Ghana, 11–14
　integration, barriers to, 88–89
　necessity for, 78–80
　pico-hydro, 172
　resources, 76
　sources, 1–2
　sustainability indicators, 82
　sustainable development, 81–82
　in West Africa, 82–84
Renewable Energy Act, 89–90, 168–169
Renewable Energy Association of the Gambia (REAGAM), 180
Renewable Energy Bill, 179–180
Renewable Energy for Rural Economic Development (RERED), 57*t*
Renewable Energy Law, 183
Renewable Energy Master Plan, Nigeria, 172
Renewable Energy Plan (2010-2020), 180
Renewable Energy Technologies (RETs), 90–91
Reservoirs, hydropower plants
　African Union Commission, 205
　climate change, 202
　infrastructure, 202–203
　life-cycle stages, 204
　operation and maintenance, 203
　preventing failure, 204–205
　water-energy nexus, 204

Resettlement
  Bui Dam, 125–133, 128*f*
  involuntary, 117
  Jama Resettlement, 125–126, 130*f*,
    131–132, 132*f*
Resettlement Planning Framework (RPF), 125
Rio Declaration on Environment and
    Development, 163
River Blindness, 130–131
Riverine ecosystems, impoundments impact
    on, 113–114
Rural and Renewable Energy Agency, 176–177
Rural Electrification Agency, 174–175, 184
Rural Energy Development Programme
    (REDP), 54, 57*t*

S
Sahel, 97
Scaling-Up Renewable Energy Program
    (SREP), 181–182
Second National Environmental Action
    Plan, 181
Seismic-related failure, 201–202
Senegal River, 98
Senegal's Electricity Law, 183
Senegal's National Adaptation Programme of
    Action, 184
Shared socioeconomic pathways (SSPs), 45
Sierra Leone legislation, 175–176
Small- and medium-scale enterprises (SMEs),
    2–3
Small-Scale Hydropower Program (SSHPP),
    167–168
Snowy Mountains Eng. Corp (SMEC),
    112–113
Societal change, Sub-Saharan Africa, 69–71
Solar Farm Project, 2, 50-MW, 55–56
Specialist Advisor, 31
Statistical Product for Service Solution
    (SPSS), 143
Stockholm Declaration, 159–160, 162
Strategic Climate Fund (SCF), 42–43
Sub-Saharan Africa (SSA), 137–138
  climate change, 63–66
  economy, 63–65
  electrical energy supply in, 66–69
  electricity demand in, 69*f*
  energy supply sector, 66
  floods and droughts, 66
  greenhouse gases, 66
  hydraulic potential, 67*f*

hydropower energy generation, 63–66,
    69–71
  installed hydropower capacity, 68*t*
  technical capacities, 65
  water resources, 64–65
Sunbelt Potential of Photovoltaics, 83
Sustainable development (SD), 43
  climate change mitigation, 79–80
  commercial energy consumption, 78*f*
  energy consumption, 75–76
  indicators for renewable energy, 82
  installed generation capacity, 79*f*
  management of hydropower projects, 53–54
  necessity for, 78–80
  renewable energy and, 81–82
  scheme, 81*f*
Sustainable Development Goals (SDGs),
    78–80, 164–165
Sustainable Energy for All initiative
    (SE4ALL), 164
Swift-spinning shaft, 16

T
Technical potential, in West Africa, 83,
    97–99, 100*t*
Theoretical potential, in West Africa, 82
Tiger Leaping Gorge Dam, 56, 57*t*
Tilapias, 110
Tipaimukh Dam, 56, 57*t*
Trongsa Micro Hydel Project, 57*t*

U
United Nations Conference on
    Environment and Development
    (UNCED), 159–160, 163
United Nations Conference on the Human
    Environment in Stockholm, 159–160
United Nations Development Programme
    (UNDP), 54, 172
United Nations Economic Commission for
    Africa (UNECA), 175
United Nations Environment Program (UNEP),
    160–161
United Nations Framework Convention on
    Climate Change (UNFCCC), 65–66
United Nations General Assembly, 164
United Nations Sustainable Energy for All
    (UNSE4ALL), 161
University of Energy and Natural Resources
    (UENR) in Ghana, 31

V

Volta Aluminum Company (VALCO), 6
Volta River Authority's (VRA) gas power
    plant, 2–3
Volta River, hydropower potential, 99
Volume of water, 3

W

Wa municipality, 137, 139–140, 141*f*
    climate variability/change
        adaptation, 151–155, 152*t*
        chi-square tests, 150*t*, 153*t*
        local people's perception, 146–151,
            148*f*, 149*t*
        water insecurity, 154–155
    CWSA, 140–141
    data analysis, 143
    data collection tools, 142–143
    map, 141*f*
    research design, 140–143
    sampling procedure, 140–142, 142*t*
    water insufficiency, economic activities,
        143–146, 144*f*, 145*t*
Waste management, Bui Dam, 131–132
Water-energy nexus, 204
Water insecurity, climate variability, 154–155
Water insufficiency, economic activities,
    143–146
Water resources, 96*f*
Wave energy, climate change and, 14*f*

Wave Energy Conversion Devices (WECDs),
    12–13, 15–23
Wave motion, 16
WCD. *See* World Commission on Dams (WCD)
Wells' turbine, 16
West Africa, 96
    biofuel potential in, 87*t*
    biomass potential for energy, 85–86
    feed-in tariff, 91*t*
    geographic and climatic conditions, 76–77
    layout, 97
    map of, 77*f*
    observations, 89–90
    policy recommendations, 90–91
    RE integration, barriers to, 88–89
    renewable energy potential in, 82–84
    solar irradiance of, 84*t*
    solar resource map, 85*f*
    water resources, 96*f*
    wind energy potentials in, 85, 86*t*, 87*f*
West African Power Pool (WAPP), 161
    ECOWAS, 165, 167–168
    Liberia's membership, 177–178
    Sierra Leone, 176
White Paper, ECOWAS, 167
Wind energy, 81–82, 85, 86*t*
Working Groups (WGs), 161
World Bank, 102–103, 117
World Commission on Dams (WCD), 121–122
World Meteorological Organization (WMO),
    160–161

Printed in the United States
By Bookmasters